THEORIES
AND
OBSERVATION
IN SCIENCE

CENTRAL ISSUES IN PHILOSOPHY SERIES

BARUCH A. BRODY
series editor

~~~~~~~~~~~~~~~~~~~~~~~~~~~~~~~~~~~~

Baruch A. Brody
MORAL RULES AND PARTICULAR CIRCUMSTANCES

Hugo A. Bedau
JUSTICE AND EQUALITY

Mark Levensky
HUMAN FACTUAL KNOWLEDGE

George I. Mavrodes
THE RATIONALITY OF BELIEF IN GOD

Robert Sleigh
NECESSARY TRUTH

David M. Rosenthal
MATERIALISM AND THE MIND-BODY PROBLEM

Richard Grandy
THEORIES AND OBSERVATION IN SCIENCE

Gerald Dworkin
DETERMINISM, FREE WILL, AND MORAL RESPONSIBILITY

David P. Gauthier
MORALITY AND RATIONAL SELF-INTEREST

Charles Landesman
THE FOUNDATONS OF KNOWLEDGE

Adrienne and Keith Lehrer
THEORY OF MEANING

edited by

RICHARD E. GRANDY
Princeton University

THEORIES

AND

OBSERVATION

IN SCIENCE

Prentice-Hall, Inc., Englewood Cliffs, New Jersey

Library of Congress Catalog Card Number: 72–5316

Printed in the United States of America

ISBN: C 0–13–913400–X
P 0–13–913392–5

10 9 8 7 6 5 4 3 2 1

PRENTICE-HALL INTERNATIONAL, INC., London
PRENTICE-HALL OF AUSTRALIA, PTY. LTD., Sydney
PRENTICE-HALL OF CANADA, LTD., Toronto
PRENTICE-HALL OF INDIA PRIVATE LIMITED, New Delhi
PRENTICE-HALL OF JAPAN, INC., Tokyo

Foreword

The Central Issues in Philosophy series is based upon the conviction that the best way to teach philosophy to introductory students is to experience or to *do* philosophy with them. The basic unit of philosophical investigation is the particular problem, and not the area or the historical figure. Therefore, this series consists of sets of readings organised around well-defined, manageable problems. All other things being equal, problems that are of interest and relevance to the student have been chosen.

Each volume contains an introduction that clearly defines the problem and sets out the alternative positions that have been taken. The selections are chosen and arranged in such a way as to take the student through the dialectic of the problem; each reading, besides presenting a particular point of view, criticizes the points of view set out earlier.

Although no attempt has been made to introduce the student in a systematic way to the history of philosophy, classical selections relevant to the development of the problem have been included. As a side benefit, the student will therefore come to see the continuity, as well as the breaks, between classical and contemporary thought. But in no case will a selection be included merely for its historical significance; clarity of expression and systematic significance are the main criteria for selection.

BARUCH A. BRODY

Contents

INTRODUCTORY ESSAY   **1**

ERNST MACH
The Economical Nature of Physics   16

NORMAN CAMPBELL
Definition of a Theory   22

RUDOLPH CARNAP
Testability and Meaning   27

R. B. BRAITHWAITE
*The Nature of Theoretical Concepts
and the Role of Models in an Advanced Science*   47

C. G. HEMPEL
*The Empiricist Criteria of
Cognitive Significance:
Problems and Changes*   53

ISRAEL SCHEFFLER
*Prospects of
a Modest Empiricism, I*   73

J. J. C. SMART
*The Reality of
Theoretical Entities*   93

GROVER MAXWELL

Theories, Frameworks,
and Ontology   *104*

HILARY PUTNAM

What Theories Are Not   *111*

RICHARD C. JEFFREY

Review of Putnam   *124*

NORWOOD R. HANSON

Observation   *129*

PAUL K. FEYERABEND

On the Interpretation
of Scientific Theories   *147*

W. V. O. QUINE

Posits and Reality   *154*

PETER ACHINSTEIN

On Meaning-Dependence   *162*

PAUL K. FEYERABEND

On the "Meaning" of Scientific Terms   *176*

BIBLIOGRAPHICAL ESSAY.   *184*

Introduction

THE DISTINCTIONS

A distinctive feature of scientific theories is that they contain expressions such as 'electron', 'magnetic field', and 'DNA molecule', terms which do not occur in everyday discourse. Not only are the terms foreign to our usual conversations, but the items they designate are unfamiliar to nonscientists. This feature of scientific theories has, not surprisingly, attracted the attention of philosophers of science and has led them to draw a distinction between two classes of terms and objects. However, neither the task of drawing the distinction nor that of explaining its significance has proved as easy as was originally thought; it is unusual for any two philosophers to agree either on what the distinction is or on its importance. The readings in this anthology have been selected with the intention of tracing the continuing debate about the nature of the distinction as well as the more recent discussions concerning whether or not a meaningful distinction can even be drawn. Probably one reason that the difficulty of drawing the distinction in a satisfactory manner was underestimated is that philosophers have focused primarily on the more exciting disagreements about the consequences of the distinction. Thus disagreements about the nature of the distinction have been masked by the assumption that all of the writers were attempting to describe the same dichotomy.

In characterizing the feature of scientific theories that attracted attention, we mentioned both scientific terms and the objects they designate. Corresponding to the difference between the terms and the objects there are two basic ways of drawing the distinction. The terminological distinction is usually made by calling terms that are peculiar to scientific theories 'theoretical terms' and the others 'ob-

1

servational terms.' When the distinction is drawn between the two kinds of objects rather than the two kinds of terms, the labels are usually 'observable' and 'unobservable.' The contrast between observable objects and unobservable objects is roughly intended to be that between those objects of which we can have some direct perceptual experience and those which we can perceive only indirectly.

The connection between the ontological and terminological distinctions is a complex one. The distinction between observable and unobservable objects was thought to be important because it was the difference between objects which are directly experienced and those which are not. The terminological distinction is at first blush a linguistic recasting of the same difference. According to one definition, observational terms are characterized by the fact that we are directly acquainted with the objects to which they refer. In the case of unobservable entities, there are two kinds of possible philosophical concern: one might raise the question as to whether such entities exist, or the question of how we come to know anything about them. For theoretical terms, however, one might also raise questions about how they are learned or whether they are meaningful.

Despite the connection between the motivations for the two distinctions, they are not obviously coextensive. If all and only observational terms applied to observable entities and all and only theoretical terms applied to unobservable entities, then the two distinctions would coincide. It seems plausible, however, that in any definition of 'theoretical', theoretical terms apply to what are intuitively thought of as observable objects. People contain protein molecules, pieces of iron generate magnetic fields, and so on. It also seems plausible that observable predicates such as 'six-legged' and 'brown' are applicable, for example, to insects too small to see with the naked eye. The article by Putnam and the review by Jeffrey present more detailed arguments to show the divergence between the two distinctions. This divergence, while it does show that earlier writers who did not recognize the difference between the two distinctions were confused, does not show that both distinctions are not tenable.

THE ONTOLOGICAL DISTINCTION

Lack of clarity as to exactly what is being explicated has considerably influenced the discussion of the problems. For example, the original emphasis on the distinction between kinds of objects and the continued conflation of the ontological and terminological distinctions has led to an emphasis on theoretical terms which refer to spatiotemporal objects, such as electrons or DNA molecules. This seems to distort the situation, since theoretical terms are by no means confined to terms which function as collective nouns designating spatio-temporal objects. There are terms such as 'magnetic field' or 'electromagnetic wave' which can be said to apply to objects only if one includes fields and waves in the category of objects. Another large class of theoretical terms designates quantities or properties of objects or systems of objects; for example, something described as having a mass of n grams, having potential energy E, or being ionized. Most of these properties can be properties of observable or unobservable objects indiscriminately. Hempel * has pointed out that one cannot claim that there are two senses of mass, one theoretical and one observational; the mass of a volume of gas, which is observable, is the sum of the masses of its molecules, which are unobservable. Even some of the theoretical terms which function as collective nouns such as 'electrical discharge' or 'mass of H_2O with temperature less than zero degrees Centigrade' are applicable to such eminently observable things as lightning bolts and ice cubes.

The contrast we noted in the beginning between the kinds of objects which are observable and those which are introduced in theoretical discourse has led some philosophers to consider the question of whether or not theoretical entities exist. Some philosophers, such as Mach, in the first selection, have argued that since we have no experience of unobservable entities, they can only be convenient fictions which are introduced to simplify our theories. There are, however, arguments against the tenability of any distinction between observable and unobservable objects. The explanation of an unobservable entity appeals to the difference between observing an object directly and merely observing its effects. But if one considers the fact that seeing any object involves photons reflected from (or emitted by) the object impinging on the retina of the observer, the notion of directly observing begins to lose its intuitive clarity. There seems to be only a slight difference of degree between directly seeing and observing through a magnifying glass, and only a slight difference between using a magnifying glass and using a microscope or telescope. This argument, elaborated at length in the selections by Maxwell and Smart, leads to the conclusion that there are only differences in degree between the observability of various kinds of objects. But if observability is merely a matter of degree, then there seems to be no plausible way of drawing a sharp line on this basis between objects which do and objects which do not exist. Under the influence of these considerations, most philosophers have given up the attempt to distinguish observables from unobservables on this basis and focus on the terminological distinction.

* Hempel, C. G., "On the 'Standard Conception' of Scientific Theories," Minnesota Studies in the Philosophy of Science, Vol. IV, ed. Herbert Feigl et al. (Minneapolis: University of Minnesota Press, 1970).

THE TERMINOLOGICAL DISTINCTION

This introduction began by remarking that one striking feature of scientific theories is that they contain theoretical terms; if one also reflects on the fact that science is but a refinement and continuation of common-sense exploration of the world around us, it would seem that the distinction has interest in areas other than theoretical science. For example, one could examine nonscientific terms and try to distinguish those having the direct sort of connection with experience that observational terms do from those that do not. And thus an understanding of the way in which theoretical terms derive their meaning from observational discourse would be enlightening about nonscientific language as well. One important application of the analysis of the relationship between the kinds of terms might be to examine various terms of everyday language to determine whether they are meaningful at all. That is, once the connection between observational terms and theoretical terms has been analyzed and we know what relation the theoretical terms must bear to observation, we could decide whether various other terms bore the same kind of relation. For example, philosophers of a particular skeptical bent have often wondered whether various key terms in theological or ethical disputes have any meaning. If one could show on the basis of the analysis of the language of science that these crucial terms were meaningless, many fruitless debates could be avoided.

The terminological distinction not only gives rise to problems about theoretical terms but also influences other areas of philosophy of science. One important example is that if one takes the dichotomy seriously, then theories should be viewed as having three distinct parts to their structure: those statements that contain only observational terms, those that contain both theoretical and observational terms, and those that contain only theoretical terms. If the distinction is untenable or unimportant, it is likely that a different analysis of theory structure would be more illuminating.

A second example is the influence the distinction has on the analysis of how scientific statements are confirmed or refuted. There are many unsolved problems concerning when sentences in a given vocabulary confirm, inductively support, or refute other sentences in the same vocabulary. For example, to what extent does a given set of observed meltings of copper samples confirm an assertion about the melting point of copper? * If one thinks that the basic data of science consist of statements in observational vocabulary, then there is a further

---

* For discussion of some of the problems see C. G. Hempel, "Studies in the Logic of Confirmation," *Aspects of Scientific Explanation* (New York: The Free Press, 1965), and R. Carnap, *Logical Foundations of Probability* (Chicago: University of Chicago Press, 1950, rev. 1962).

problem of analyzing how statements in observational vocabulary can support statements which contain theoretical terms.

A third example of the influence of the distinction is that it lends support to the conception that the aim of science is to give a systematic and economical description of observable phenomena. This conception in turn shapes our conception of scientific progress. If the aim of science is description and prediction of observable phenomena, then progress in science can only be measured by its descriptive and predictive success. Progress must be a linear and cumulative matter, moreover, since the observational vocabulary and hence the range of phenomena to be systematized are fixed. This conception tends to undervalue or ignore the role of theories in designing experiments which can open up whole new areas for investigation. For example, theories about atomic and nuclear structure have led to the creation of artificial elements.*

The most important problems generated by the distinction concern, of course, the nature of theoretical terms. The reason for calling the nontheoretical terms observational is that it has been assumed (or argued) that terms such as 'purple', 'apple', and so on can be used to describe observations. This assumption is closely tied to a view of language according to which the meaning of a word consists in the rules that connect the word with the experiences of the speaker of the language. Given this assumption, one seems to be faced with the problem that the theoretical terms are not connected in any direct way with experience and hence it is unclear whether they have any meaning at all. For example, if one thinks that the meaning of 'apple' consists of the rules correlating the word with sensations of touch, sight, or taste, then it is difficult to see what meaning 'electron' has, since electrons cannot be tasted, seen, or touched. This line of argument leads to the conclusion that it is questionable if theoretical terms are cognitively significant, which is the linguistic formulation of the question of whether or not theoretical entities exist.

Various versions of this kind of argument have led philosophers to attribute an inferior semantic status to theoretical terms. The degree and nature of the inferiority depends on what auxiliary philosophical hypotheses about language, experience, and reality are invoked. If one believes that meaning consists entirely in direct connection with experience through semantic rules and that the theoretical terms are meaningful, then one is led to the position illustrated in the excerpt from Mach that theoretical terms must be explicitly definable using only observational terms. That is, for any theoretical term we can find

---

* See T. S. Kuhn, *The Structure of Scientific Revolutions* (Chicago: University of Chicago Press, 1962), for a discussion of the distorting influence the linear view of scientific progress has had on the history of science.

an equivalent expression that contains only observational terms. All attempts to support this position by actually supplying the explicit definitions have failed, however, and there are convincing reasons for thinking that this program cannot be carried out. The reasons for this failure are discussed in the selections by Carnap, Hempel, and Scheffler.

The most extreme position concerning the inferiority of theoretical terms, *Fictionalism,* denies that theoretical terms play any essential role whatsoever. According to this position the terms are not explicitly definable but can be eliminated in principle. It is considered to be a historical or psychological accident that such terms have been used in science. A defense of this position would require a method for reconstructing scientific theories in general in such a way that theoretical terms would never appear. Two methods based on technical results in mathematical logic have been proposed for achieving this goal, but both seem to have serious defects. Fictionalism is discussed at some length in the selection by Scheffler and references are given there to the required technical elaborations.

A slightly less negative view of theoretical terms is to treat them as meaningless in the strict sense, but useful nevertheless. This position is usually labeled *Instrumentalism* since it relies heavily on the analogy between theoretical terms and tools or instruments. The usual formulation of the position is that theoretical terms serve as useful instruments in organizing observational statements but have no cognitive significance. A prerequisite for defending this position would be to explain how expressions can be useful though meaningless. In practice, defenders of this position tend to slip into either the previous position and treat theoretical terms as dispensable or else attribute some partial significance to the terms, a position we will consider in a moment. For example, if an instrumentalist attempts to explain that theoretical terms are useful for establishing inductive or deductive connections between observation statements, then it seems that he is merely refusing to apply the predicate 'meaningful' to the terms though attributing to them all the usual properties of meaningful terms. If the terms were actually meaningless, then any sentence in which they occur should be nonsensical and could hardly be used to establish the relevant connections. One variation on this theme is to interpret statements containing theoretical terms as 'inference tickets'; that is, to reinterpret statements of the form "If A, then B" as a rule which permits the inference of B from A. The distinction between rules of a formal system and its statements is an arbitrary one, however. Any statement can be turned into a rule and any nonlogical rule can be turned into a statement.

In its mildest form, the attribution of inferior status to theoretical terms consists of considering such terms to be partially meaningful.

According to this partial interpretation position, theoretical terms obtain partial significance via their connections with the observational terms. This position requires considerable elaboration before it can be evaluated; in particular, some explanation must be given of how the partial significance accrues, and an analysis must be given of the connection between the two kinds of terms. The greatest bulk of the literature on the terminological problem has taken the form of making precise the partial interpretation position.

The earliest adherent of this position, Campbell, was concerned almost solely with partial interpretation as a problem in the philosophy of science. In the 1920s and 30s Carnap, Schlick, and other members of the Vienna circle attempted to use the partial interpretation view to support more general philosophical theses. As we mentioned earlier, they wished to demonstrate that many traditional philosophical problems were pseudo-problems because the terms used in posing and solving the problems were meaningless. The basic viewpoint behind this was similar to the one discussed at the beginning of this introduction: observational terms are straightforwardly meaningful and any other terms must derive their meaning, if any, from connections with observational terms. Thus a solution to the problem of what kind of connection exists between observational terms and legitimate theoretical terms in science was expected to provide a criterion for demarcating the boundary between sense and nonsense in general. Once the nature of the connection between legitimate theoretical terms and observation was specified, it could be shown (it was hoped) that many metaphysical concepts were not so connected with experience and hence lacked cognitive significance.*

The difficulties in explicating the connection between the two kinds of terms, however, proved to be considerably greater than anticipated. A number of definitions of 'partial interpretation' were offered, all of which failed in one respect or another. They were either too strict and thus ruled out as meaningless theoretical terms which seem essential for scientific practice, or else they were too loose and demonstrably let in any term whatsoever. The selection from Campbell gives one of the earliest characterizations of partial interpretation; Carnap and Braithwaite offer expositions and defenses of the partial interpretation approach. The survey by Hempel reviews the more important attempts to clarify the concept of partial interpretation and offers some suggestions for further research. The selection from Putnam, in addition

---

* The history of this movement is discussed in J. Jorgenson, "The Development of Logical Empiricism," Vol. II, No. 9 of the *International Encyclopedia of Unified Science* (Chicago: University of Chicago Press, 1951), and *The Legacy of Logical Positivism*, S. Barker and P. Achinstein, eds. (Baltimore: Johns Hopkins Press, 1969).

to attacking the distinctions involved, offers reasons for thinking that the entire conception of partial interpretation is ill-conceived.

The partial interpretation approach to theoretical terms is closely associated with a particular view of theories. Theories are often considered to consist of an abstract uninterpreted calculus plus rules relating the uninterpreted symbols to observational terms. The symbolic calculus would consist of equations or statement forms such as $ax = y$ and the connecting rules would specify, for example, that if two bodies balance on a scale, then the x value for one is the same as the x value for the other. The portion of the theory consisting of theoretical terms is treated as an uninterpreted calculus because these terms are viewed as meaningless until they are connected with the observational terms. There is ambiguity, however, in the notion of partial interpretation. Sometimes partial interpretation is thought of as statements in the scientific language itself that connect observational and theoretical statements. For example, one such connection might be contained in the statement: "Temperature-average kinetic energy of the molecules of the substance." At other times partial interpretation is thought of as semantic rules in a second (meta-) language in which one speaks about the first language. The paradigm of such a rule would be that "red" means red. The approach via semantic rules is more natural if one takes seriously the notion that the theoretical terms are uninterpreted symbols; one naturally thinks of explaining (interpreting) an equation by saying that 'F' stands for force, 'x' for mass, and so forth. But the actual statements that one gives to interpret the symbols must either contain theoretical terms in the second language or take the form of scientific assertions such as the one above about temperature. If the sentences which are to correlate the observables with uninterpreted symbols are thought of as part of the scientific language itself, then it is difficult to see how they can be fully meaningful, since they contain uninterpreted symbols.

THE NEW VIEW OF THEORY AND OBSERVATION

One insight that has emerged, particularly in the Hempel paper, from the efforts expended on definitions of partial interpretation is that attention should be focused not on individual terms in isolation but rather on the whole set of terms of a theory and their relation to each other. This is a shift in emphasis away from the definability of individual terms and toward the systematic connections between terms in context. One reason that this insight emerged so slowly is that the partial interpretation approach developed out of the explicit definability position of Mach. On this particular point, Campbell seems to have had a better appreciation of the situation than many later writers. He explicitly considers the possibility that it is only combinations of theoretical notions that are related to observational situations and that individual theoretical notions cannot be so related.

This possibility arises because the basic problem has been formulated on the assumption of a theory of meaning which makes the meaningfulness of the theoretical terms dubious. If this theory that views meaning as specified by the rules directly correlating expressions with bits of the world can be discredited, then the entire problem about the distinction needs to be reformulated. That is, if understanding a language is not a matter of correlating individual terms with specific kinds of experiences but rather a matter of confronting the totality of experience with the total conceptual structure of the language, then it becomes plausible that all terms are in a sense theoretical. Another way of explaining the respect in which observation is theoretical concerns the fact that what an observer reports depends upon his anticipations, and these anticipations will usually depend upon his theoretical knowledge. The selection from Hanson illustrates the way in which observers' reports are influenced almost as much by what they expect to see as by what is before them.

A less abstruse way of arriving at a similar conclusion about the vagueness of the distinction is to reflect on the epistemological certainty that was thought to be attached to observational terms. The intuitive notion of an observational term seems to be that any individual who understands the term can correctly and easily decide whether or not the term applies in a given situation without recourse to instruments or inferences based on theoretical premises. For example, one can decide whether an object is purple simply by looking. Of course one cannot *always* decide correctly whether the term applies or not, since the object in question may be too far away, or the light may be inadequate or peculiar. These considerations lead to refinements of the explanation of observationality that can take either of two forms: one can explain that the observational property is one that can be correctly and easily applied under standard conditions, or else one can take the observational property to be not simply 'purple' but rather 'looks purple'. The first position is difficult to expound without using theoretical information in deciding what are and what are not standard cases. The second position departs from the original intuition motivating the distinction, since the connection between 'purple' and 'looks purple' poses many of the same problems as the connection between theoretical and observational terms. But if theoretical knowledge is required to determine what are observation terms, then the notion seems to lack the epistemological solidity that was initially so appealing. The second position is of considerable interest and has generated a significant amount of discussion in the philosophical literature, but it is beyond the scope of this volume. The distinction is now analogous to the original observational-theoretical dichotomy, but it does not arise in the context of scientific theory. Another consideration that can be adduced in support of the holistic conception of language is the following: Anyone who can learn to use the term 'red' can also be taught to say "emits or reflects electromagnetic radiation of wavelengths between 5800 and 6200 angstroms," since they apply to the same objects. This may seem an unfair example, since one can only know that the two are coextensive by knowing and thus understanding a considerable amount of theory and by using the relevant instruments. Otherwise one would not even understand the theoretical terms used in the second phrase. The most serious difficulty in drawing the distinction sharply is to make precise what is meant by 'instrument' and 'knowing some theory'. For example, is 'magnet' an observational or a theoretical term? One can test whether an object is a magnet by bringing a piece of iron near it; does the piece of iron count as an instrument? Does knowing that magnets attract iron objects count as knowing a theory? These considerations and similar ones are expounded in the articles by Quine and Feyerabend in systematic arguments to show that language has the type of holistic structure sketched above.

Recently some writers have admitted that exactly where one draws the line between observational and theoretical terms is arbitrary, but they still claim that the difference is an important one. Such claims can either be confused ways of persisting in attacking the same old problem without meeting the argument that the problem is based on a mistaken theory of language, or they can be slightly misleading ways of reformulating the problem. One can recognize that there is no sharp absolute distinction that can be drawn for the whole of science and is unchanging over time without giving up the original insight that there is something worthy of investigation. If one is studying the early stages of the history of science during which the phenomena of magnetism are not yet understood, then it makes sense to take 'magnet' as a theoretical term. If one is considering some more advanced stage at which the basic magnetic phenomena are understood and some other theory is being developed, then it makes sense to treat 'magnet' as an observational term.

The basic question can be reformulated as follows: There is no absolute distinction between observational and theoretical terms, but at any given stage of scientific development, or in the reconstruction of such a stage, there are some terms which are taken as understood and others which are newly introduced. Statements in this accepted vocabulary are treated as unproblematic from the viewpoint of the theory in question and in general such statements will be capable of being readily decided by a scientist who knows the theories that are being assumed. The problem is to explain the manner in which the new theoretical terms—that is, those peculiar to the new theory being developed—acquire their meaning.

This kind of relativism which makes the distinction relative to theories under consideration should be distinguished from the kind of breakdown which the object dichotomy suffered. The terminological distinction presents us with layers of theoreticalness which are fairly clearcut even though complexly interwoven; the dichotomy between objects, on the other hand, provides only a continuum with no particularly distinguished structure.

Something analogous to the partial interpretation position is the usual answer suggested to this problem, and it runs afoul of some of the same objections and difficulties discussed above. If one thinks of terms as being either meaningful or meaningless, then it is difficult to see how a connection between the antecedently meaningless theoretical terms and the antecedently meaningful terms can be established. The terminology of meaning has proved a slippery one, however, and the problem can also be formulated in terms of *understanding* the new theoretical concepts. It seems intuitively plausible that a word is either meaningful or meaningless, but the concept of understanding a word is intuitively less of an all-or-nothing affair. 'Knowing what a word means' is a vague notion which admits of great variation. In particular, one may know perfectly well what a word means without being able to define it explicitly.

Thus, if we approach the problem of establishing connections between old and new terms from the perspective of understanding how one learns the theoretical terms, the difficulties dissolve. It is a familiar occurrence that one can encounter a number of sentences containing a word which is initially unfamiliar, and from a sufficiently large sample of sentences can learn to use the word correctly. Learning to use theoretical terms such as 'electron' might thus be construed as learning to manipulate the equations and statements containing the terms. The equations and statements will include both some couched entirely in the novel terms and some which contain terms familiar from previous theories.

As mentioned earlier, the problem is closely tied to questions in the philosophy of language. The matter could be phrased in such a way as to make it seem impossible to solve: before one can understand theoretical terms one must understand the theory in which they occur, but before the theory can be understood, the vocabulary it contains must be understood. Such a bootstrap operation seems more plausible, however, if one does not insist on comprehension being an all-or-nothing affair. For example, someone who learns to play chess cannot be said to understand how the game is played until he understands such terms as 'rook', 'king', 'check', and so forth; but he cannot understand the significance of such terms unless he understands the game. The king is not simply a piece of wood of a particular shape, it is a piece which has a particular role in the game. What in fact happens is that the person does not learn one word at a time but rather learns the entire set of words and their relations and implications at the same time. This holistic view of language and meaning makes sense of what actually occurs in a way the earlier theory could not.

This perspective on meaning has led Ryle to suggest the terminology of 'theory-laden' rather than 'theoretical'. Terms such as 'electron' and 'rook' are said to be laden with the atomic theory or the rules of chess. The motivation for the terminology is that to know what an electron is includes knowing the relevant physical theories; and knowing what a rook is involves knowing how chess is played. This new position seems to resolve the old problem, but it confronts a new problem peculiar to it. If understanding the terms of a theory involves knowing the theory, and if the meaning of a term is to be thought of as the role the term plays in the theory, then it might be argued that 'electron' means different things in different theories about electrons. But if this is the case, then it is difficult to see how any two theories can be rivals, since they have no common vocabulary and no common subject matter. Two theories which appear to disagree about the properties of electrons only apparently disagree, since they attribute different meanings to 'electron'. This is intuitively undesirable, since we want to say that Bohr's theory of the atom was accepted because it was a better theory of the atom than its predecessors; that is, it did a better job of describing the same objects its predecessors tried to portray. This difficulty, under the label 'incommensurability', is discussed in the selections by Feyerabend, who welcomes incommensurability as a fundamental feature of science rather than as a difficulty; however his tolerance of the situation is not widely shared.*

The selection by Achinstein attacks the concept of theory-ladenness by arguing that there are a number of different concepts involved and that they have not been sufficiently distinguished. He argues further that the various senses are important for different reasons and that their conflation can only lead to confusion. A term might be theory-dependent in any of a number of ways: it might require appeal to a theory T to define the term—we might need the principles of T to devise experiments to measure the values of the quantity that the term

---

* Kuhn, in *Structure of Scientific Revolutions*, has agreed with Feyerabend in some respects, but prefers the term 'partial communication' for describing the relationship between different theories. "Reflections on My Critics," in *Criticism and the Growth of Knowledge*, Imre Lakatos and Alan Musgrave, eds. (New York: Cambridge University Press, 1970), and the revised version of Kuhn's *Structure of Scientific Revolutions* (Chicago: University of Chicago Press), contain recent reconsiderations of Kuhn's position.

designates, or we might need the theory to determine the properties of the objects denoted by the term. Moreover, a term may be theory-laden in the sense of being used only in one theory, or it may be theory-laden in that the statements about it are known on the basis of a particular theory but the term is used in many other theories.

One conclusion that appears inescapable in view of the history of the problem of theoretical terms is that the problem cannot be adequately formulated, let alone solved, without more careful examination of actual theories. For example, we mentioned above that consideration of actual theories shows that theoretical terms often designate quantities (mass, energy), processes (emission of electrons, magnetization), or any number of other creatures (fields, orbits, waves) quite unlike objects in the usual sense. There is no *a priori* reason to think that terms that are diverse in these respects are similar with respect to the way they are interpreted. It might prove to be the case that the development of quantitative terms consists partly in extending the range of values of the quantity and partly in identifying the quantities with other quantities. For example, the development of the concept of temperature has been partly due to the creation of instruments to measure higher and lower temperatures and to the revision of laws involving temperatures when these laws are extended to low or high temperatures, and partly in the identification of the macroscopic concept of temperature with the microscopic concept of average kinetic energy of the molecules. The first kind of development should be more easily analyzed than the second. Or if a term for a particular process is first introduced in a new theory, it might still be the case that the entities involved in the process are familiar from previous theories, and again the problem of interpretation seems less formidable.

Another important point that has not been mentioned thus far is that once one formulates the problem on the basis of a relative distinction between antecedently understood and antecedently not understood terms, there are far fewer problematic terms. Consideration of theories in historical perspective shows that any particular theory usually adds only slightly to the theoretical vocabulary, while most of the terms are taken over relatively unchanged from previous theories. This is to be expected, since the motivation for creating a new theory is usually to explain some range of phenomena which an earlier theory treated less successfully.

Consider, for example, Bohr's theory of the atom. Bohr's theory consisted of four basic postulates:

(1) The atom consists of electrons moving about the nucleus under the effect of electrical (Couloumb) attraction between electron and nucleus and obeying the classical laws applicable to charged bodies in motion.

(2) Only certain of the infinitely many classically permissible orbits can be realized, those being the orbits for which the angular momentum is an integral multiple of Planck's constant.

(3) Contrary to classical electromagnetic theory, the rotating electron does not radiate energy except when it moves from one orbit to another.

(4) The change of orbits is discontinuous and the energy given off determines the frequency of the electromagnetic radiation emitted.

Bohr's theory was proposed to solve several problems which earlier theories of atomic structure had been unable to handle. Earlier theories did not explain why energy was radiated in discrete quantities; they did not explain how the atom could be stable since, according to classical theory, the electron should continuously radiate energy until the orbit degenerated. The first of these problems is not so much solved as dismissed by Bohr's theory, but the theory did permit accurate predictions of what wavelengths would be emitted by various atoms. Not only did it permit accurate quantitative predictions but it also provided a derivation of Balmer's formula governing spectral lines, which had been established by experiment but not connected with any previous theory.

Our present concern is not with the confirmation of the theory, however, but with its vocabulary. The concepts of energy, velocity, position, mass, change, orbit, angular momentum, electrical attractive force, and electromagnetic radiation are all taken from earlier mechanical and electromagnetic theory. The only differences are that they are now being applied to somewhat smaller dimensions than those usually considered and that the second and third postulates deny the applicability of some of the classical principles to some of the processes. Other classical principles are still applied in the calculation of the energy, angular momentum, and so forth, and the derivation of quantitative predictions from the theory depended essentially on these classical connections. The notions of electron and nucleus had been used previously in other atomic theories and some of their important properties such as the values of mass and charge were known. Planck's constant had already appeared in his original theory of electromagnetic radiation of heated bodies and in Einstein's analysis of the photoelectric effect. Thus the only completely new concept in Bohr's theory is that of the discontinuous transition of an electron from one orbit to another. The most puzzling feature of the theory is not the newly introduced term but the peculiar restrictions placed on the applicability of classical theories.

This example, even so superficially analyzed, gives a hint of the kind of problem and solution that has taken the place of our original problem. The fact that there are different kinds of theoretical terms which receive interpretation in various ways need not move us to despair. The problem is not so easily stated nor so monolithic as we originally thought, but the solution does not seem so difficult. The recognition that there are different types of theoretical terms and that they receive interpretation in diverse ways, both in a particular theory and in relation to other theories, may ultimately prove fruitful. One advantage is that we are now encouraged to pay more attention to theories in their natural habitat and are less dependent on a preconceived theory of meaning. It seems likely that the analysis of the more detailed structure of theories will prove more rewarding than an extremely general characterization that applies to all theories and in any case, the detailed analysis is a necessary prerequisite to any accurate generalizations.

The Economical Nature
of Physics

Physics is experience, arranged in economical order. By this order not only is a broad and comprehensive view of what we have rendered possible, but also the defects and the needful alterations are made manifest, exactly as in a well-kept household. Physics shares with mathematics the advantages of succinct description and of brief, compendious definition, which precludes confusion, even in ideas where, with no apparent burdening of the brain, hosts of others are contained. Of these ideas the rich contents can be produced at any moment and displayed in their full perceptual light. Think of the swarm of well-ordered notions pent up in the idea of the potential. Is it wonderful that ideas containing so much finished labor should be easy to work with?

Our first knowledge, thus, is a product of the economy of self-preservation. By communication, the experience of *many* persons, individually acquired at first, is collected in *one*. The communication of knowledge and the necessity which every one feels of managing his stock of experience with the least expenditure of thought, compel us to put our knowledge in economical forms. But here we have a clue which strips science of all its mystery, and shows us what its power really is. With respect to specific results it yields us nothing that we could not reach in a sufficiently long time without methods. There is no problem in all mathematics that cannot be solved by direct counting. But with the present implements of mathematics many operations of counting can be performed in a few minutes which without mathematical methods would take a lifetime. Just as a single human being, restricted wholly to the fruits of his own labor, could never amass a

Excerpted from pp. 197–206 of *Popular Scientific Lectures* by kind permission of the publishers The Open Court Publishing Co., La Salle, Illinois.

fortune, but on the contrary the accumulation of the labor of many men in the hands of one is the foundation of wealth and power, so, also, no knowledge worthy of the name can be gathered up in a single human mind limited to the span of a human life and gifted only with finite powers, except by the most exquisite economy of thought and by the careful amassment of the economically ordered experience of thousands of co-workers. What strikes us here as the fruits of sorcery are simply the rewards of excellent housekeeping, as are the like results in civil life. But the business of science has this advantage over every other enterprise, that from *its* amassment of wealth no one suffers the least loss. This, too, is its blessing, its freeing and saving power.

The recognition of the economical character of science will now help us, perhaps, to understand better certain physical notions.

Those elements of an event which we call "cause and effect" are certain salient features of it, which are important for its mental reproduction. Their importance wanes and the attention is transferred to fresh characters the moment the event or experience in question becomes familiar. If the connexion of such features strikes us as a necessary one, it is simply because the interpolation of certain intermediate links with which we are very familiar, and which possess, therefore, higher authority for us, is often attended with success in our explanations. That *ready* experience fixed in the mosaic of the mind with which we meet new events, Kant calls an innate concept of the understanding (*Verstandesbegriff*).

The grandest principles of physics, resolved into their elements, differ in no wise from the descriptive principles of the natural historian. The question, "Why?" which is always appropriate where the explanation of a contradiction is concerned, like all proper habitudes of thought, can overreach itself and be asked where nothing remains to be understood. Suppose we were to attribute to nature the property of producing like effects in like circumstances; just these like circumstances we should not know how to find. Nature exists once only. Our schematic mental imitation alone produces like events. Only in the mind, therefore, does the mutual dependence of certain features exist.

All our efforts to mirror the world in thought would be futile if we found nothing permanent in the varied changes of things. It is this that impels us to form the notion of substance, the source of which is not different from that of the modern ideas relative to the conservation of energy. The history of physics furnishes numerous examples of this impulse in almost all fields, and pretty examples of it may be traced back to the nursery. "Where does the light go to when it is put out?" asks the child. The sudden shrivelling up of a hydrogen balloon is inexplicable to a child; it looks everywhere for the large body which was just there but is now gone.

Where does heat come from? Where does heat go to? Such childish questions in the mouths of mature men shape the character of a century.

In mentally separating a body from the changeable environment in which it moves, what we really do is to extricate a group of sensations on which our thoughts are fastened and which is of relatively greater stability than the others, from the stream of all our sensations. Absolutely unalterable this group is not. Now this, now that member of it appears and disappears, or is altered. In its full identity it never recurs. Yet the sum of its constant elements as compared with the sum of its changeable ones, especially if we consider the continuous character of the transition, is always so great that for the purpose in hand the former usually appear sufficient to determine the body's identity. But because we can separate from the group every single member without the body's ceasing to be for us the same, we are easily led to believe that after abstracting all the members something additional would remain. It thus comes to pass that we form the notion of a substance distinct from its attributes, of a thing-in-itself, whilst our sensations are regarded merely as symbols or indications of the properties of this thing-in-itself. But it would be much better to say that bodies or things are compendious mental symbols for groups of sensations—symbols that do not exist outside of thought. Thus, the merchant regards the labels of his boxes merely as indexes of their contents, and not the contrary. He invests their contents, not their labels, with real value. The same economy which induces us to analyse a group and to establish special signs for its component parts, parts which also go to make up other groups, may likewise induce us to mark out by some single symbol a whole group.

On the old Egyptian monuments we see objects represented which do not reproduce a single visual impression, but are composed of various impressions. The heads and the legs of the figures appear in profile, the head-dress and the breast are seen from the front, and so on. We have here, so to speak, a mean view of the objects, in forming which the sculptor has retained what he deemed essential, and neglected what he thought indifferent. We have living exemplifications of the processes put into stone on the walls of these old temples, in the drawings of our children, and we also observe a faithful analogue of them in the formation of ideas in our own minds. Only in virtue of some such facility of view as that indicated, are we allowed to speak of a body. When we speak of a cube with trimmed corners—a figure which is not a cube— we do so from a natural instinct of economy, which prefers to add to an old familiar conception a correction instead of forming an entirely new one. This is the process of all judgment.

The crude notion of "body" can no more stand the test of analysis than can the art of the Egyptians or that of our little children. The physicist who sees a body flexed, stretched, melted, and vaporized, cuts up this body into smaller permanent parts; the chemist splits it up into elements. Yet even an element is not unalterable. Take sodium. When warmed, the white, silvery mass becomes a liquid, which, when the heat is increased and the air shut out, is transformed into a violet vapor, and on the heat being still more increased glows with a yellow light. If the name sodium is still retained, it is because of the continuous character of the transitions and from a necessary instinct of economy. By condensing the vapor, the white metal may be made to reappear. Indeed, even after the metal is thrown into water and has passed into sodium hydroxide, the vanished properties may by skillful treatment still be made to appear; just as a moving body which has passed behind a column and is lost to view for a moment may make its appearance after a time. It is unquestionably very convenient always to have ready the name and thought for a group of properties wherever that group by any possibility can appear. But more than a compendious economical symbol for these phenomena, that name and thought is not. It would be a mere empty word for one in whom it did not awaken a large group of well-ordered sense-impressions. And the same is true of the molecules and atoms into which the chemical element is still further analysed.

True, it is customary to regard the conservation of weight, or, more precisely, the conservation of mass, as a direct proof of the constancy of matter. But this proof is dissolved, when we go to the bottom of it, into such a multitude of instrumental and intellectual operations, that in a sense it will be found to constitute simply an equation which our ideas in imitating facts have to satisfy. That obscure, mysterious lump which we involuntarily add in thought, we seek for in vain outside the mind.

It is always, thus, the crude notion of substance that is slipping unnoticed into science, proving itself constantly insufficient, and ever under the necessity of being reduced to smaller and smaller world-particles. Here, as elsewhere, the lower stage is not rendered indispensable by the higher which is built upon it, no more than the simplest mode of locomotion, walking, is rendered superfluous by the most elaborate means of transportation. Body, as a compound of light and touch sensations, knit together by sensations of space, must be as familiar to the physicist who seeks it, as to the animal who hunts its prey. But the student of the theory of knowledge, like the geologist and the astronomer, must be permitted to reason back from the forms which are created before his eyes to others which he finds ready made for him.

All physical ideas and principles are succinct directions, frequently involving subordinate directions, for the employment of economically classified experiences, ready for use. Their conciseness, as also the fact that their contents are rarely exhibited in full, often invests them with the semblance of independent existence. Poetical myths regarding such ideas—for example, that of Time, the producer and devourer of all things—do not concern us here. We need only remind the reader that even Newton speaks of an *absolute* time independent of all phenomena, and of an absolute space—views which even Kant did not shake off, and which are often seriously entertained to-day. For the natural inquirer, determinations of time are merely abbreviated statements of the dependence of one event upon another, and nothing more. When we say the acceleration of a freely falling body is 9.810 metres per second, we mean the velocity of the body with respect to the centre of the earth is 9.810 metres greater when the earth has performed an additional 86400th part of its rotation—a fact which itself can be determined only by the earth's relation to other heavenly bodies. Again, in velocity is contained simply a relation of the position of a body to the position of the earth.[1] Instead of referring events to the earth we may refer them to a clock, or even to our internal sensation of time. Now, because all are connected, and each may be made the measure of the rest, the illusion easily arises that time has significance independently of all.[2]

The aim of research is the discovery of the equations which subsist between the elements of phenomena. The equation of an ellipse expresses the universal *conceivable* relation between its co-ordinates, of which only the real values have *geometrical* significance. Similarly, the equations between the elements of *phenomena* express a universal, mathematically conceivable relation. Here, however, for many values only certain directions of change are *physically* admissible. As in the ellipse only certain *values* satisfying the equation are realised, so in the physical world only certain *changes* of value occur. Bodies are always accelerated towards the earth. Differences of temperature, left to themselves, always grow less; and so on. Similarly, with respect to space, mathematical and physiological researches have shown that the space of experience is simply an *actual* case of many conceivable cases, about whose peculiar properties experience alone can instruct us. The elucidation which this idea diffuses cannot be questioned, despite the absurd uses to which it has been put.

Let us endeavor now to summarise the results of our survey. In the economical schematism of science lie both its strength and its weakness. Facts are always represented at a sacrifice of completeness and never

---

[1] It is clear from this that all so-called elementary (differential) laws involve a relation to the Whole.

[2] If it be objected, that in the case of perturbations of the velocity of rotation of the earth, we could be sensible of such perturbations, and being obliged to have some measure of time, we should resort to the period of vibration of the waves of sodium light—all that this would show is that for practical reasons we should select that event which best served us as the *simplest* common measure of the others.

with greater precision than fits the needs of the moment. The incongruence between thought and experience, therefore, will continue to subsist as long as the two pursue their course side by side; but it will be continually diminished.

In reality, the point involved is always the completion of some partial experience; the derivation of one portion of a phenomenon from some other. In this act our ideas must be based directly upon sensations. We call this measuring.[3] The condition of science, both in its origin and in its application, is a *great relative stability* of our environment. What it teaches us is interdependence. Absolute forecasts, consequently, have no significance in science. With great changes in celestial space we should lose our co-ordinate systems of space and time.

When a geometer wishes to understand the form of a curve, he first resolves it into small rectilinear elements. In doing this, however, he is fully aware that these elements are only provisional and arbitrary devices for comprehending in parts what he cannot comprehend as a whole. When the law of the curve is found he no longer thinks of the elements. Similarly, it would not become physical science to see in its self-created, changeable, economical tools, molecules and atoms, realities behind phenomena, forgetful of the lately acquired sapience of her older sister, philosophy, in substituting a mechanical mythology for the old animistic or metaphysical scheme, and thus creating no end of suppositious problems. The atom must remain a tool for representing phenomena, like the functions of mathematics. Gradually, however, as the intellect, by contact with its subject-matter, grows in discipline, physical science will give up its mosaic play with stones and will seek out the boundaries and forms of the bed in which the living stream of phenomena flows. The goal which it has set itself is the *simplest* and *most economical* abstract expression of facts.

•　　•　　•　　•　　•　　•　　•　　•　　•　　•　　•　　•　　•

[3] Measurement, in fact, is the definition of one phenomenon by another (standard) phenomenon.

NORMAN R. CAMPBELL

Definition of a Theory

A theory is a connected set of propositions which are divided into two groups. One group consists of statements about some collection of ideas which are characteristic of the theory; the other group consists of statements of the relation between these ideas and some other ideas of a different nature. The first group will be termed collectively the "hypothesis" of the theory; the second group the "dictionary". The hypothesis is so called, in accordance with the sense that has just been stated, because the propositions composing it are incapable of proof or of disproof by themselves; they must be significant, but, taken apart from the dictionary, they appear arbitrary assumptions. They may be considered accordingly as providing a "definition by postulate" of the ideas which are characteristic of the hypothesis. The ideas which are related by means of the dictionary to the ideas of the hypothesis are, on the other hand, such that something is known about them apart from the theory. It must be possible to determine, apart from all knowledge of the theory, whether certain propositions involving these ideas are true or false. The dictionary relates some of these propositions of which the truth or falsity is known to certain propositions involving the hypothetical ideas by stating that if the first set of propositions is true then the second set is true and vice versa; this relation may be expressed by the statement that the first set implies the second.

In scientific theories (for it seems that there may be sets of propositions having exactly the same features in departments of knowledge other than science) the ideas connected by means of the dictionary to the hypothetical ideas are always concepts . . . that is collections of fundamental judgements related in laws by uniform association; and

Reprinted from *Foundations of Science* by Norman R. Campbell, pp. 122–25 by kind permission of the publishers, The Cambridge University Press.

the propositions involving these ideas, of which the truth or falsity is known, are always laws. Accordingly those ideas involved in a theory which are not hypothetical ideas will be termed concepts; it must be remembered that this term is used in a very special sense; concepts depend for their validity on laws, and any proposition in which concepts are related to concepts is again a law. Whether there is any necessary limitation on the nature of the ideas which can be admitted as hypothetical ideas is a question which requires much consideration; but one limitation is obviously imposed at the outset by the proviso that propositions concerning them are arbitrary, namely that they must not be concepts. As a matter of fact the hypothetical ideas of most of the important theories of physics, but not of other sciences, are mathematical constants and variables. (Except when the distinction is important, the term "variable" will be used in this chapter to include constants.)

The theory is said to be true if propositions concerning the hypothetical ideas, deduced from the hypothesis, are found, according to the dictionary, to imply propositions concerning the concepts which are true, that is to imply laws; for all true propositions concerning concepts are laws. And the theory is said to explain certain laws if it is these laws which are implied by the propositions concerning the hypothetical ideas.

An illustration will make the matter clearer. To spare the feelings of the scientific reader and to save myself from his indignation, I will explain at the outset that the example is wholly fantastic, and that a theory of this nature would not be of the slightest importance in science. But when it has been considered we shall be in a better position to understand why it is so utterly unimportant, and in what respects it differs from valuable scientific theories.

The hypothesis consists of the following mathematical propositions:

(1) $u, v, w, \ldots$ are independent variables.
(2) a is a constant for all values of these variables.
(3) b is a constant for all values of these variables.
(4) $c = d$, where c and d are dependent variables.

The dictionary consists of the following propositions:

(1) The assertion that $(c^2 + d^2)a = R$, where R is a positive and rational number, implies the assertion that the resistance of some definite piece of pure metal is R.

(2) The assertion that $cd/b = T$ implies the assertion that the temperature of the same piece of pure metal is T.

From the hypothesis we deduce

$$(c^2 + d^2)a \left/ \frac{cd}{b} \right. = 2ab = \text{constant}.$$

Interpreting this proposition by means of the dictionary we arrive at the following law:

The ratio of the resistance of a piece of pure metal to its absolute temperature is constant.

This proposition is a true law (or for our purpose may be taken as such). The theory is therefore true and explains the law.

This example, absurd though it may seem, will serve to illustrate some of the features which are of importance in actual theories. In the first place, we may observe the nature of the propositions, involving respectively the hypothetical ideas and the concepts, which are stated by the dictionary to imply each other. When the hypothetical ideas are mathematical variables, the concepts are measurable concepts (an idea of which much will be said hereafter), and the propositions related by mutual implication connect the variables, or some function of them, to the same number as these measurable concepts. When such a relation is stated by the dictionary it will be said for brevity that the function of the hypothetical ideas "is" the measurable concept; thus, we shall say that $(c^2 + d^2)a$ and cd/b "are" respectively the resistance and temperature. But it must be insisted that this nomenclature is adopted only for brevity; it is not meant that in any other sense of that extremely versatile word "is" $(c^2 + d^2)a$ is the resistance; for there are some senses of that word in which a function of variables can no more "be" a measurable concept than a railway engine can "be" the year represented by the same number.

If an hypothetical idea is directly stated by the dictionary to be some measurable concept, that idea is completely determined and every proposition about its value can be tested by experiment. But in the example which has been taken this condition is not fulfilled. It is only functions of the hypothetical ideas which are measurable concepts. Moreover since only two functions, which involve four mathematical variables and between which one relation is stated by the hypothesis, are stated to be measurable concepts, it is impossible by a determination of those concepts to assign definitely numerical values to them. If some third function of them had been stated to be some third measurable concept, then it would have been possible to assign to all of them numerical values in a unique manner. If further some fourth function has been similarly involved in the dictionary, the question would have arisen whether the values determined from one set of three functions is consistent with those determined from another set of three.

These distinctions are important. There is obviously a great difference between a theory in which some proposition based on experiment can be asserted about each of the hypothetical ideas, and one in which nothing can be said about these ideas separately, but only about combinations of them. There is also a difference between those in which several statements about those ideas can be definitely shown to be consistent and those in which such statements are merely known not to be inconsistent. In these respects actual theories differ in almost all possible degrees; it very often happens that some of the hypothetical ideas can be directly determined by experiment while others cannot; and in such cases there is an important difference between the two classes of ideas. Those which can be directly determined are often confused with the concepts to which they are directly related, while those which cannot are recognized as distinctly theoretical. But it must be noticed that a distinction of this nature has no foundation. The ideas of the hypothesis are never actually concepts; they are related to concepts only by means of the dictionary. Whatever the nature of the dictionary, all theories have this in common that no proposition based on experimental evidence can be asserted concerning the hypothetical ideas except on the assumption that the propositions of the theory are true. This is a most important matter which must be carefully borne in mind in all our discussions.

It will be observed that in our example there are no propositions in the dictionary relating any of the independent variables of the hypothesis to measurable concepts. This feature is characteristic of such theories. The nature of the connection between the independent variables and the concepts is clear from the use made, in the deduction of the laws, of the fact that a and b are constants, not varying with the independent variables. The conclusion that the electrical resistance is proportional to the absolute temperature would not follow unless $(c^2 + d^2)a$ were the resistance in the same state of the system the same as that in which cd/b is the temperature; and on the other hand it would not follow if a and b were not the same constants in all the propositions of the dictionary. Accordingly the assertion that a or b is a constant must imply that it is the same so long as the state of the system to which the concepts refer is the same; the independent variables on the contrary may change without a corresponding change in the state of the system. If therefore there is to be in the dictionary a proposition introducing the independent variables, it must state that a change in the independent variables does *not* imply a change in the state of the system; the omission of these variables from the dictionary must be taken to mean a definite negative statement. On the other hand, the independent variables may bear some relation to measurable concepts, so long as these concepts are not properties of the system. Thus, in almost all theories of this type, one of the independent variables is called the "time", and the use of this name indicates that it is related in some manner to the physically measurable "time" since some agreed datum. What exactly is this relation we shall have to inquire in the third part of this volume, but it is to be noted that a relation between one of the independent variables and physically measured time is not inconsistent with the statement that a change in this variable does not imply any change in the state of the system; for it is one of the essential properties of a system that its state should be, in a certain degree and within certain limits, independent of time.

In some theories again, there are dependent variables which are not mentioned in the dictionary. But in such cases the absence of mention is not to be taken as involving the definite assertion that there is no relation between these variables and the concepts. It must always be regarded as possible that a further development of the theory may lead to their introduction into the dictionary.

RUDOLF CARNAP

Testability and Meaning

INTRODUCTION

1. Our Problem: Confirmation, Testing and Meaning

Two chief problems of the theory of knowledge are the question
of meaning and the question of verification. The first question asks
under what conditions a sentence has meaning, in the sense of cognitive, factual meaning. The second one asks how we get to know
something, how we can find out whether a given sentence is true or
false. The second question presupposes the first one. Obviously we
must understand a sentence, that is we must know its meaning, before
we can try to find out whether it is true or not. But, from the point
of view of empiricism, there is a still closer connection between the
two problems. In a certain sense, there is only one answer to the two
questions. If we knew what it would be for a given sentence to be
found true then we would know what its meaning is. And if for two
sentences the conditions under which we would have to take them
as true are the same, then they have the same meaning. Thus the
meaning of a sentence is in a certain sense identical with the way we
determine its truth or falsehood; and a sentence has meaning only if
such a determination is possible.

If by verification is meant a definitive and final establishment of
truth, then no (synthetic) sentence is ever verifiable, as we shall see.*

Reprinted from Rudolf Carnap, "Testability and Meaning," *Philosophy of
Science*, Vol. 3 (1936), 419–71, and Vol. 4 (1937), 1–40, by kind permission of the
author and publisher. Copyrights 1936, 1937, The Williams & Wilkins Company,
Baltimore, Maryland.

* A synthetic sentence is one which cannot be shown to be true or false on
logical or mathematical grounds alone. See p. 34. —Ed.

We can only confirm a sentence more and more. Therefore we shall speak of the problem of *confirmation* rather than of the problem of verification. We distinguish the *testing* of a sentence from its confirmation, thereby understanding a procedure—for example, the carrying out of certain experiments—which leads to a confirmation in some degree either of the sentence itself or of its negation. We shall call a sentence *testable* if we know such a method of testing for it; and we call it *confirmable* if we know under what conditions the sentence would be confirmed. As we shall see, a sentence may be confirmable without being testable; for example, if we know that our observation of such and such a course of events would confirm the sentence, and such and such a different course would confirm its negation without knowing how to set up either this or that observation.

2. The Older Requirement of Verifiability

The connection between meaning and confirmation has sometimes been formulated by the thesis that a sentence is meaningful if and only if it is verifiable, and that its meaning is the method of its verification. The historical merit of this thesis was that it called attention to the close connection between the meaning of a sentence and the way it is confirmed. This formulation thereby helped, on the one hand, to analyze the factual content of scientific sentences, and, on the other hand, to show that the sentences of trans-empirical metaphysics have no cognitive meaning. But from our present point of view, this formulation, although acceptable as a first approximation, is not quite correct. By its oversimplification, it led to a too narrow restriction of scientific language, excluding not only metaphysical sentences but also certain scientific sentences having factual meaning. Our present task could therefore be formulated as that of a modification of the requirement of verifiability. It is a question of a modification, not of an entire rejection of that requirement. For among empiricists there seems to be full agreement that at least some more or less close relation exists between the meaning of a sentence and the way in which we may come to a verification or at least a confirmation of it.

The requirement of verifiability was first stated by Wittgenstein,[1] and its meaning and consequences were exhibited in the earlier pub-

---

[1] Wittgenstein [1].

[2] I use this geographical designation because of lack of a suitable name for the movement itself represented by this Circle. It has sometimes been called Logical Positivism, but I am afraid this name suggests too close a dependence upon the older Positivists, especially Comte and Mach. We have indeed been influenced to a considerable degree by the historical positivism, especially in the earlier stage of our development. But today we would like a more general name for our movement,

lications of our *Vienna Circle*;[2] it is still held by the more conservative wing of this Circle.[3] The thesis needs both explanation and modification. What is meant by 'verifiability' must be said more clearly. And then the thesis must be modified and transformed in a certain direction.

Objections from various sides have been raised against the requirement mentioned not only by anti-empiricist metaphysicians but also by some empiricists, for example by Reichenbach,[4] Popper,[5] Lewis,[6] Nagel,[7] and Stace.[8] I believe that these criticisms are right in several respects; but on the other hand, their formulations must also be modified. The theory of confirmation and testing which will be explained in the following chapters is certainly far from being an entirely satisfactory solution. However, by more exact formulation of the problem, it seems to me, we are led to a greater convergence with the views of the authors mentioned and with related views of other empiricist authors and groups. The points of agreement and of still existing differences will be evident from the following explanations.

A first attempt at a more detailed explanation of the thesis of verifiability has been made by Schlick[9] in his reply to Lewis' criticisms. Since 'verifiability' means 'possibility of verification' we have to answer two questions: (1) what is meant in this connection by 'possibility'? and (2) what is meant by 'verification'? Schlick—in his explanation of 'verifiability'—answers the first question, but not the second one. In his answer to the question: what is meant by 'verifiability of a sentence S', he substitutes the fact described by S for the process of verifying S. Thus he thinks, for example, that the sentence S_1: "Rivers flow up-hill," is verifiable, because it is logically possible that rivers flow up-hill. I agree with him that this fact is logically possible and that the sentence S_1 mentioned above is verifiable—or, rather, confirmable, as we prefer to say for reasons to be explained soon. But I think his reasoning which leads to this result is not quite correct. S_1 is confirmable, not because of the logical possibility of the fact de-

---

comprehending the groups in other countries which have developed related views (see: Congress [1], [2]). The term *'Scientific Empiricism'* (proposed by *Morris* [1] p. 285) is perhaps suitable. In some historical remarks in the following, concerned chiefly with our original group I shall however use the term 'Vienna Circle'.

[3] Schlick [1] p. 150, and [4]; Waismann [1] p. 229.

[4] Reichenbach [1] and earlier publications; [3].

[5] Popper [1].

[6] Lewis [2] has given the most detailed analysis and criticism of the requirement of verifiability.

[7] Nagel [1].

[8] Stace [1].

[9] Schlick [4].

scribed in S_1, but because of the physical possibility of the process of confirmation; it is possible to test and to confirm S_1 (or its negation) by observations of rivers with the help of survey instruments.

Except for some slight differences, for example the mentioned one, I am on the whole in agreement with the views of Schlick explained in his paper. I agree with his clarification of some misunderstandings concerning positivism and so-called methodological solipsism. When I used the last term in previous publications I wished to indicate by it nothing more than the simple fact,[10] that everybody in testing any sentence empirically cannot do otherwise than refer finally to his own observations; he cannot use the results of other people's observations unless he has become acquainted with them by his own observations, for example by hearing or reading the other man's report. No scientist, as far as I know, denies this rather trivial fact. Since, however, the term 'methodological solipsism'—in spite of all explanations and warnings—is so often misunderstood, I shall prefer not to use it any longer. As to the fact intended, there is, I think, no disagreement among empiricists; the apparent differences are due only to the unfortunate term. A similar remark is perhaps true concerning the term 'autopsychic basis' ('eigenpsychische Basis').

3. Confirmation instead of Verification

If verification is understood as a complete and definitive establishment of truth then a universal sentence, for example, a so-called law of physics or biology, can never be verified, a fact which has often been remarked. Even if each single instance of the law were supposed to be verifiable, the number of instances to which the law refers—for example, the space-time-points—is infinite and therefore can never be exhausted by our observations which are always finite in number. We cannot verify the law, but we can test it by testing its single instances; that is, the particular sentences which we derive from the law and from other sentences established previously. If in the continued series of such testing experiments no negative instance is found but the number of positive instances increases then our confidence in the law will grow step by step. Thus, instead of verification, we may speak here of gradually increasing *confirmation* of the law.

Now a little reflection will lead us to the result that there is no fundamental difference between a universal sentence and a particular sentence with regard to verifiability but only a difference in degree.* Take for instance the following sentence: "There is a white sheet of paper on this table." In order to ascertain whether this thing is paper,

---

[10] Comp.: Erkenntnis 2, p. 461.

* Hempel, in "On the 'Standard Conception' of Scientific Theories," presents an argument that there is an essential difference between universal and particular sentences. —Ed.

we may make a set of simple observations and then, if there still remains some doubt, we may make some physical and chemical experiments. Here as well as in the case of the law, we try to examine sentences which we infer from the sentence in question. These inferred sentences are predictions about future observations. The number of such predictions which we can derive from the sentence given is infinite; and therefore the sentence can never be completely verified. To be sure, in many cases we reach a practically sufficient certainty after a small number of positive instances, and then we stop experimenting. But there is always the theoretical possibility of continuing the series of test-observations. Therefore here also *no complete verification is possible* but only a process of gradually increasing *confirmation*. We may, if we wish, call a sentence disconfirmed [11] in a certain degree if its negation is confirmed in that degree.

The impossibility of absolute verification has been pointed out and explained in detail by Popper.[12] In this point our present views are, it seems to me, in full accordance with Lewis[13] and Nagel.[14]

Suppose a sentence S is given, some test-observations for it have been made, and S is confirmed by them in a certain degree. Then it is a matter of practical decision whether we will consider that degree as high enough for our acceptance of S, or as low enough for our rejection of S, or as intermediate between these so that we neither accept nor reject S until further evidence will be available. Although our decision is based upon the observations made so far, nevertheless it is not uniquely determined by them. There is no general rule to determine our decision. Thus the acceptance and the rejection of a (synthetic) sentence always contains a *conventional component*. That does not mean that the decision—or, in other words, the question of truth and verification—is conventional. For, in addition to the conventional component there is always the non-conventional component—we may call it, the objective one—consisting in the observations which have been made. And it must certainly be admitted that in very many cases this objective component is present to such an overwhelming extent that the conventional component practically vanishes. For such a simple sentence, for example, "There is a white thing on this table" the degree of confirmation, after a few observations have been made, will be so high that we practically cannot help accepting the sentence. But even in this case there remains still the theoretical possibility of denying the sentence. Thus even here it is a matter of decision or convention.

[11] "Erschüttert," Neurath [6].
[12] Popper [1].
[13] Lewis [2] p. 137, note 12: "No verification of the kind of knowledge commonly stated in propositions is ever absolutely complete and final."
[14] Nagel [1] p. 144 f.

The view that no absolute verification but only gradual confirmation is possible, is sometimes formulated in this way: every sentence is a probability-sentence; for example by Reichenbach[15] and Lewis.[16] But it seems advisable to separate the two assertions. Most empiricists today will perhaps agree with the first thesis, but the second is still a matter of dispute. It presupposes the thesis that the degree of confirmation of a hypothesis can be interpreted as the degree of probability in the strict sense which this concept has in the calculus of probability; that is, as the limit of relative frequency. Reichenbach[17] holds this thesis. But so far he has not worked out such an interpretation in detail, and today it is still questionable whether it can be carried out at all. Popper[18] has explained the difficulties of such a frequency interpretation of the degree of confirmation; the chief difficulty lies in how we are to determine for a given hypothesis the series of "related" hypotheses to which the concept of frequency is to apply. It seems to me that at present it is not yet clear whether the concept of degree of confirmation can be defined satisfactorily as a quantitative concept; that is, a magnitude having numerical values. Perhaps it is preferable to define it as a merely topological concept, that is, by defining only the relations: "S_1 has the same (or, a higher) degree of confirmation than S_2 respectively," but in such a way that most of the pairs of sentences will be incomparable. We will use the concept in this way—without however defining it—only in our informal considerations which serve merely as a preparation for exact definitions of other terms. We shall later on define the concepts of complete and incomplete reducibility of confirmation as syntactical concepts, and those of complete and incomplete confirmability as descriptive concepts.

5. Some Terms and Symbols of Logic

In carrying out methodological investigations especially concerning verification, confirmation, testing etc., it is very important to distinguish clearly between logical and empirical, and so forth, psychological questions. The frequent lack of such a distinction in so-called epistemological discussions has caused a great deal of ambiguity and misunderstanding. In order to make quite clear the meaning and nature of our definitions and explanations, we will separate the two kinds of definitions. . . . We shall define concepts belonging to logic,

---

[15] Reichenbach [1].
[16] Lewis [2] p. 133.
[17] Reichenbach [2] p. 271 ff.; [3] p. 154 ff.
[18] Popper [1] Chapter VIII; for the conventional nature of the problem compare my remark in "Erkenntnis" Vol. 5, p. 292.

or more precisely, to logical syntax, although the choice of the concepts to be defined and of the way in which they are defined is suggested in some respects by a consideration of empirical questions—as is often the case in laying down logical definitions. The logical concepts defined here will be applied later on . . . in defining concepts of an empirical analysis of confirmation. These descriptive, that is, non-logical, concepts belong to the field of biology and psychology, namely to the theory of the use of language as a special kind of human activity.

In the following logical analysis we shall make use of some few *terms of logical syntax,* which may here be explained briefly.[19] The terms refer to a language-system, say L, which is supposed to be given by a system of rules of the following two kinds. The formative rules state how to construct sentences of L out of the symbols of L.* The transformative rules† state how to deduce a sentence from a class of sentences, the so-called premisses, and which sentences are to be taken as true unconditionally, that is, without reference to premisses. The transformative rules are divided into those which have a logico-mathematical nature; they are called logical rules or L-rules (this 'L-' has nothing to do with the name 'L' of the language); and those of an empirical nature, for example physical or biological laws stated as postulates; they are called physical rules or P-rules.

We shall take here 'S', 'S_1', 'S_2' and so forth as designations of sentences (not as abbreviations for sentences). We use '$\sim S$' as designation of the negation of S. (Thus, in this connection, '$\sim$' is not a symbol of negation but a syntactical symbol, an abbreviation for the words 'the negation of'.) If a sentence S can be deduced from the sentences of a class C according to the rules of L, S is called a *consequence* of C; and moreover an L-consequence, if the L-rules are sufficient for the deduction, otherwise a P-consequence. S_1 and S_2 are called *equipollent‡* (with each other) if each is a consequence of the other. If S can be shown to be true on the basis of the rules of L, S is called *valid* in L; and moreover L-valid or *analytic,* if true on the basis of the L-rules alone, otherwise P-valid. If, by application of the rules of L, S can be shown to be false, S is called *contravalid;* and L-contravalid or *contradictory,* if by L-rules alone, otherwise P-contravalid. If S is neither valid nor contravalid S is called *indeterminate.* If S is

---

[19] For more exact explanations of these terms see Carnap [4]; some of them are explained also in [5].

* These are usually now called the formation rules and the sentences are the well-formed sequences of symbols. —Ed.

† The transformative rules are now more commonly called the rules of inference. —Ed.

‡ Sentences S_1 and S_2, which are logically deducible from one another, are also said to be logically equivalent. —Ed.

neither analytic nor contradictory, in other words, if its truth or false-hood cannot be determined by logic alone, but needs reference either to P-rules or to the facts outside of language, S is called *synthetic*. Thus the totality of the sentences of L is classified in the following way:

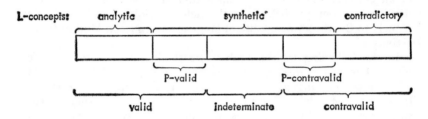

A sentence S_1 is called incompatible with S_2 (or with a class C of sentences), if the negation $\sim S_1$ is a consequence of S_2 (or of C, respectively). The sentences of a class are called mutually independent if none of them is a consequence of, or incompatible with, any other of them.

The most important kind of predicates occurring in a language of science is that of the predicates attributed to space-time-points (or to small space-time-regions). For the sake of simplicity we shall restrict the following considerations—so far as they deal with predicates—to those of this kind. The attribution of a certain value of a physical function, for example, of temperature, to a certain space-time-point can obviously also be expressed by a predicate of this kind. The following considerations, applied here to such predicates only, can easily be extended to descriptive terms of any other kind.

In order to be able to formulate examples in a simple and exact way we will use the following symbols. We take 'a', 'b', and so forth, as names of space-time-points (or of small space-time-regions), that is as abbreviations for quadruples of space-time-coordinates; we call them *individual constants*. 'x', 'y', etc. will be used as corresponding variables; we will call them *individual variables*. We shall use 'P', 'P_1', 'P_2' and so forth, and 'Q', 'Q_1' and so forth as *predicates;* if no other indication is given, they are supposed to be predicates of the kind described. The sentence '$Q_1(b)$' is to mean: "The space-time-point b has the property Q_1." * Such a sentence consisting of a predicate followed by one or several individual constants as arguments, will be called a *full sentence* of that predicate.

* Although Carnap confines his attention to space-time points and properties of space-time points, the logical notation is not so restricted. One could take the individual constants as names of persons and the predicates to be predicates applicable to persons, for example. —Ed.

Connective symbols: '∼' for 'not' (negation), 'V' for 'or' (disjunction), '·' for 'and' (conjunction), '⊃' for 'if – then' (implication),* '≡' for 'if – then –, and if not – then not –' (equivalence).† '∼$Q(a)$' is the negation of a full sentence of 'Q'; it is sometimes also called a full sentence of the predicate '∼Q'.

Operators‡ '$(x)P(x)$' is to mean: "every point has the property P" (*universal* sentence; the first '(x)' is called the universal *operator,* and the sentential function '$P(x)$' its *operand*). '$(\exists x)P(x)$' is to mean: "There is at least one point having the property P" (*existential* sentence; '$(\exists x)$' is called the existential operator and '$P(x)$' its operand). (In what follows, we shall not make use of any other operators than universal and existential operators with individual variables, as described here.) In our later examples we shall use the following abbreviated notation for universal sentences of a certain form occurring very frequently. If the sentence '$(x)[- - -]$' is such that '$- - -$' consists of several partial sentences which are connected by '∼', 'V' and so forth and each of which consists of a predicate with 'x' as argument, we allow omission of the operator and the arguments. Thus, for example, instead of '$(x)\ (P_1(x) \supset P_2(x))$' we shall write shortly '$P_1 \supset P_2$'; and instead of '$(x)\ [Q_1(x) \supset (Q_3(x) \equiv Q_2(x))]$' simply '$Q_1 \supset (Q_3 \equiv Q_2)$'. The form '$P_1 \supset P_2$' is that of the simplest physical laws; it means: "If any space-time-point has the property P_1, it has also the property P_2."

7. Definitions

By an (explicit) definition of a descriptive predicate 'Q' with one argument we understand a sentence of the form

$(D:)$ $$Q(x) \equiv \ldots x \ldots$$

where at the place of '$\ldots x \ldots$' a sentential function—called the *definiens*—stands which contain 'x' as the only free variable.§ For several arguments the form is analogous. We will say that a definition D is based upon the class C of predicates if every descriptive symbol

---

* The logical notation PQ does not always correspond exactly to the informal 'if – then' locution. PQ can be defined as v P v Q so that PQ is true whenever P is false or Q true. In the ordinary use of 'if – then' in many contexts there is an implicit assumption that the truth of P is relevant to the truth of Q. Some of the difficulty of translating 'if – then' locution as PQ will be discussed on pp. 36–37. —Ed.

† Equivalence also corresponds to the locution 'if and only if'. —Ed.

‡ The more common term now is quantifiers; (x) is the *universal quantifier* and $(\exists x)$ or (Ex) the *existential quantifier*. —Ed.

§ A variable is free in a sentence form if there is no quantifier governing the variable. —Ed.

occurring in the definiens of D belongs to C. If the predicates of a class C are available in our language we may introduce other predicates by a chain of definitions of such a kind that each definition is based upon C and the predicates defined by previous definitions of the chain.

Definition 9. A definition is said to have atomic (or molecular, or generalized, or essentially generalized) form, if its definiens has atomic (or molecular, or generalized, or essentially generalized, respectively) form.

Theorem 5. If 'P' is defined by a definition D based upon C, 'P' is reducible to C. If D has molecular form, 'P' is completely reducible to C. If D has essentially generalized form, 'P' is incompletely reducible to C.

Proof. 'P' may be defined by '$P(x) \equiv \ldots x \ldots$'. Then, for any b, '$P(b)$' is equipollent to '$\ldots b \ldots$' and hence in the case of molecular form, according to Theorem 2, completely reducible to C, and in the other case, according to Theorems 3 and 4, reducible to C.

Let us consider the question whether the so-called *disposition-concepts* can be defined, that is, predicates which enunciate the disposition of a point or body for reacting in such and such a way to such and such conditions, for example 'visible', 'smellable', 'fragile', 'tearable', 'soluble', 'indissoluble' etc. We shall see that such disposition-terms cannot be defined by means of the terms by which these conditions and reactions are described, but they can be introduced by sentences of another form. Suppose, we wish to introduce the predicate 'Q_3' meaning "soluble in water." Suppose further, that 'Q_1' and 'Q_2' are already defined in such a way that '$Q_1(x, t)$' means "the body x is placed into water at the time t," and '$Q_2(x, t)$' means "the body x dissolves at the time t." Then one might perhaps think that we could define 'soluble in water' in the following way: "x is soluble in water" is to mean "whenever x is put into water, x dissolves," in symbols:

$$(D:) \qquad Q_3(x) \equiv (t)[Q_1(x, t) \supset Q_2(x, t)].$$

But this definition would not give the intended meaning of 'Q_3'. For, suppose that c is a certain match which I completely burnt yesterday. As the match was made of wood, I can rightly assert that it was not soluble in water; hence the sentence '$Q_3(c)$' (S_1) which asserts that the match c is soluble in water, is false. But if we assume the definition D, S_1 becomes equipollent with '$(t) [Q_1(c, t) \supset Q_2(c, t)]$' ($S_2$). Now the match c has never been placed and on the hypothesis made never can be so placed. Thus any sentence of the form '$Q_1(c, t)$' is false for any value of 't'. Hence S_2 is true, and, because of D, S_1 also is true, in contradiction to the intended meaning of S_1.* 'Q_3' cannot be defined by

---

* It is important to recall that $P \supset Q$ is true whenever P is false. —Ed.

D, nor by any other definition. But we can introduce it by the following sentence:

$$(R:) \qquad (x)(t)[Q_1(x, t) \, P \supset (Q_3(x) \equiv Q_2(x, t))],$$

in words: "if any thing x is put into water at any time t, then, if x is soluble in water, x dissolves at the time t, and if x is not soluble in water, it does not." This sentence belongs to that kind of sentences which we shall call reduction sentences.

8. Reduction Sentences

Suppose, we wish to introduce a new predicate 'Q_3' into our language and state for this purpose a pair of sentences of the following form:

$$(R_1) \qquad\qquad Q_1 \supset (Q_2 \supset Q_3)$$
$$(R_2) \qquad\qquad Q_4 \supset (Q_5 \supset \sim Q_3)$$

Here, 'Q_1' and 'Q_4' may describe experimental conditions which we have to fulfill in order to find out whether or not a certain space-time-point b has the property Q_3, i.e., whether '$Q_3(b)$' or '$\sim Q_3(b)$' is true. 'Q_2' and 'Q_5' may describe possible results of the experiments. Then R_1 means: if we realize the experimental condition Q then, if we find the result Q_2, the point has the property Q_3. By the help of R_1, from '$Q_1(b)$' and '$Q_2(b)$', '$Q_3(b)$' follows. R_2 means: if we satisfy the condition Q_4 and then Q_5, the point has not the property Q_3. By the help of R_2, from '$Q_4(b)$' and '$Q_5(b)$', '$\sim Q_3(b)$' follows. We see that the sentences R_1 and R_2 tell us how we may determine whether or not the predicate 'Q_3' is to be attributed to a certain point, provided we are able to determine whether or not the four predicates 'Q_1', 'Q_2', 'Q_4', and 'Q_5' are to be attributed to it. By the statement of R_1 and R_2 'Q_3' is reduced in a certain sense to those four predicates; therefore we shall call R_1 and R_2 reduction sentences for 'Q_3' and '$\sim Q_3$' respectively. Such a pair of sentences will be called a reduction pair for 'Q_3'. By R_1 the property Q_3 is attributed to the points of the class $Q_1 \cdot Q_2$, by R_2 the property $\sim Q_3$ to the points of the class $Q_4 \cdot Q_5$. If by the rules of the language—either logical rules or physical laws—we can show that no point belongs to either of these classes (in other words, if the universal sentence '$\sim [(Q_1 \cdot Q_2) \lor (Q_4 \cdot Q_5)]$' is valid) then the pair of sentences does not determine Q_3 nor $\sim Q_3$ for any point and therefore does not give a reduction for the predicate Q_3. Therefore, in the definition of 'reduction pair' to be stated, we must exclude this case.

In special cases 'Q_4' coincides with 'Q_1', and 'Q_5' with '$\sim Q_2$'. In that case the reduction pair is '$Q_1 \supset (Q_2 \supset Q_3)$' and '$Q_1 \supset (\sim Q_2 \supset \sim Q_3)$'; the latter can be transformed into '$Q_1 \supset (Q_3 \supset Q_2)$'. Here the pair can be replaced by the one sentence '$Q_1 \supset (Q_3 \equiv Q_2)$' which means: if we accomplish the condition Q_1, then the point has the property Q_3 if and only if we find the result Q_2. This sentence may serve for determining the result '$Q_3(b)$' as well as for '$\sim Q_3(b)$'; we shall call it a bilateral reduction sentence. It determines Q_3 for the points of the class $Q_1 \cdot Q_2$, and $\sim Q_3$ for those of the class $Q_1 \cdot \sim Q_2$; it does not give a determination for the points of the class $\sim Q_1$. Therefore, if '$(x)(\sim Q_1(x))$' is valid, the sentence does not give any determination at all. To give an example, let '$Q'_1(b)$' mean "the point b is both heated and not heated", and '$Q''_1(b)$': "the point b is illuminated by light-rays which have a speed of 400,000 km/sec". Here for any point c, '$Q'_1(c)$' and '$Q''_1(c)$' are contravalid—the first contradictory and the second P-contravalid; therefore, '$(x)(\sim Q'_1(x))$' and '$(x)(\sim Q''_1(x))$' are valid—the first analytic and the second P-valid; in other words, the conditions Q'_1 and Q''_1 are impossible, the first logically and the second physically. In this case, a sentence of the form '$Q'_1 \supset (Q_3 \equiv Q_2)$' or '$Q''_1 \supset (Q_3 \equiv Q_2)$' would not tell us anything about how to use the predicate 'Q_1' and therefore could not be taken as a reduction sentence. These considerations lead to the following definitions.

Definition 10. **a.** A universal sentence of the form

$$(R) \qquad\qquad Q_1 \supset (Q_2 \supset Q_3)$$

is called a *reduction sentence* for 'Q_3' provided '$\sim (Q_1 \cdot Q_2)$' is not valid.

b. A pair of sentences of the forms

$$(R_1) \qquad\qquad Q_1 \supset (Q_2 \supset Q_3)$$
$$(R_2) \qquad\qquad Q_4 \supset (Q_5 \supset \sim Q_3)$$

is called a *reduction pair for* 'Q_3' provided '$\sim [(Q_1 \cdot Q_2) \vee (Q_4 \cdot Q_5)]$' is not valid.

c. A sentence of the form

$$(R_b) \qquad\qquad Q_1 \supset (Q_3 \equiv Q_2)$$

is called a *bilateral reduction sentence* for 'Q_3' provided '$(x)(\sim Q_1(x))$' is not valid.

Every statement about reduction pairs in what follows applies also to bilateral reduction sentences, because such sentences are comprehensive formulations of a special case of a reduction pair.

If a reduction pair for 'Q_3' of the form given above is valid—that is either laid down in order to introduce 'Q_3' on the basis of 'Q_1', 'Q_2', 'Q_4', and 'Q_5', or consequences of physical laws stated beforehand— then for any point c '$Q_3(c)$' is a consequence of '$Q_1(c)$' and '$Q_2(c)$', and '$\sim Q_3(c)$' is a consequence of '$Q_4(c)$' and '$Q_5(c)$'. Hence 'Q_3' is completely reducible to those four predicates.

Theorem 6. If a reduction pair for 'Q' is valid, then 'Q' is completely reducible to the four (or two, respectively) other predicates occurring.

We may distinguish between logical reduction and physical reduction, dependent upon the reduction sentence being analytic or P-valid, in the latter case for instance a valid physical law. Sometimes not only the sentence '$Q_1 \supset (Q_3 \equiv Q_2)$' is valid, but also the sentence '$Q_3 \equiv Q_2$'. (This is for example the case if '$(x)Q_1(x)$' is valid.) Then for any b, '$Q_3(b)$' can be transformed into the equipollent sentence '$Q_2(b)$', and thus 'Q_3' can be eliminated in any sentence whatever. If '$Q_3 \equiv Q_2$' is not P-valid but analytic it may be considered as an explicit definition for 'Q_3'. Thus an *explicit definition* is a special kind of a logical bilateral reduction sentence. A logical bilateral reduction sentence which does not have this simple form, but the general form '$Q_1 \supset (Q_3 \equiv Q_2)$', may be considered as a kind of conditioned definition.

If we wish to construct a language for science we have to take some descriptive (i.e. non-logical) terms as primitive terms. Further terms may then be introduced not only by explicit definitions but also by other reduction sentences. The possibility of *introduction by laws*, that is by physical reduction, is, as we shall see, very important for science, but so far not sufficiently noticed in the logical analysis of science. On the other hand the terms introduced in this way have the disadvantage that in general it is not possible to eliminate them, that is to translate a sentence containing such a term into a sentence containing previous terms only.

Let us suppose that the term 'Q_3' does not occur so far in our language, but 'Q_1', 'Q_2', 'Q_4', and 'Q_5' do occur. Suppose further that either the following reduction pair R_1, R_2 for 'Q_3':

$$(R_1) \qquad\qquad Q_1 \supset (Q_2 \supset Q_3)$$
$$(R_2) \qquad\qquad Q_4 \supset (Q_5 \supset \sim Q_3)$$

or the following bilateral reduction sentence for 'Q_3':

$$(R_b) \qquad\qquad Q_1 \supset (Q_3 \equiv Q_2)$$

is stated as valid in order to introduce 'Q_3', that is to give meaning to this new term of our language. Since, on the assumption made, 'Q_3' has no antecedent meaning, we do not assert anything about facts by the statement of R_b. This statement is not an assertion but a convention. In other words, the factual content of R_b is empty; in this respect, R_b is similar to a definition. On the other hand, the pair R_1, R_2 has a positive content. By stating it as valid, beside stating a convention concerning the use of the term 'Q_3', we assert something about facts that can be formulated in the following way without the use of 'Q_3'. If a point c had the property $Q_1 \cdot Q_2 \cdot Q_4 \cdot Q_5$, then both '$Q_3(c)$' and '$\sim Q_3(c)$' would follow. Since this is not possible for any point, the following universal sentence S which does not contain 'Q_3', and which in general is synthetic, is a consequence of R_1 and R_2:

$$(S:) \qquad \qquad \sim (Q_1 \cdot Q_2 \cdot Q_4 \cdot Q_5).$$

In the case of the bilateral reduction sentence R_b, 'Q_4' coincides with 'Q_1' and 'Q_5' with '$\sim Q_2$'. Therefore in this case S degenerates to '$\sim Q_1 \cdot Q_2 \cdot Q_1 \cdot \sim Q_2$') and hence becomes analytic. Thus a bilateral reduction sentence, in contrast to a reduction pair, has no factual content.

THE CONSTRUCTION OF A LANGUAGE-SYSTEM

17. The Problem of a Criterion of Meaning

It is not the aim of the present essay to defend the principle of empiricism against apriorism or anti-empiricist metaphysics. Taking empiricism[20] for granted, we wish to discuss, the question what is meaningful. The word 'meaning' will here be taken in its empiricist sense; an expression of language has meaning in this sense if we know how to use it in speaking about empirical facts, either actual or possible ones. Now our problem is what expressions are meaningful in this sense. We may restrict this question to sentences because expressions other than sentences are meaningful if and only if they can occur in a meaningful sentence.

Empiricists generally agree, at least in general terms, in the view that the question whether a given sentence is meaningful is closely connected with the questions of the possibility of verification, confirmation or testing of that sentence. Sometimes the two questions have been regarded as identical. I believe that this identification can be accepted only as a rough first approximation. Our real problem now is to determine the precise relation between the two questions, or generally, to state the criterion of meaning in terms of verification, confirmation or testing.

I need not emphasize that here we are concerned only with the problem of meaning as it occurs in methodology, epistemology or applied logic,[21] and not with the psychological question of meaning. We shall not consider here the questions whether any images and, if so, what images are connected with a given sentence. That these ques-

---

[20] The words 'empiricism' and 'empiricist' are here understood in their widest sense, and not in the narrower sense of traditional positivism or sensationalism or and other doctrine restricting empirical knowledge to a certain kind of experience.

[21] Our problem of meaning belongs to the field which Tarski [1] calls Semantic; this is the theory of the relations between the expressions of a language and things, properties, facts etc. described in the language.

tions belong to psychology and do not touch the methodological question of meaning, has often been emphasized.[22]

It seems to me that the question about the criterion of meaning has to be construed and formulated in a way different from that in which it is usually done. In the first place we have to notice that this problem concerns the structure of language. (In my opinion this is true for all philosophical questions, but that is beyond our present discussion.) Hence a clear formulation of the question involves reference to a certain language; the usual formulations do not contain such a reference and hence are incomplete and cannot be answered. Such a reference once made, we must above all distinguish between two main kinds of questions about meaningfulness; to the first kind belong the questions referring to a historically given language-system, to the second kind those referring to a language-system which is yet to be constructed. These two kinds of questions have an entirely different character. A question of the first kind is a theoretical one; it asks, what is the actual state of affairs; and the answer is either true or false. The second question is a practical one; it asks, how shall we proceed; and the answer is not an assertion but a proposal or decision. We shall consider the two kinds one after the other.

A *question of the first kind* refers to a given language-system L and concerns an expression E of L (i.e. a finite series of symbols of L). The question is, whether E is meaningful or not. This question can be divided into two parts: a) "Is E a sentence of L"? and b) "If so, does E fulfill the empiricist criterion of meaning"? . . . It would be advisable to avoid the terms 'meaningful' and 'meaningless' in this and in similar discussions—because these expressions involve so many rather vague philosophical associations—and to replace them by an expression of the form "a . . . sentence of L"; expressions of this form will then refer to a specified language and will contain at the place '. . .' an adjective which indicates the methodological character of the sentence, for example whether or not the sentence (and its negation) is verifiable or completely or incompletely confirmable or completely or incompletely testable and the like, according to what is intended by 'meaningful'.

18. The Construction of a Language-System L

A *question of the second kind* concerns a language-system L which is being proposed for construction. In this case the rules of L are not given, and the problem is how to choose them. We may construct L in whatever way we wish. There is no question of right or wrong, but only a practical question of convenience or inconvenience

[22] Comp. e.g. Schlick [4] p. 355.

of a system form, that is of its suitability for certain purposes. In this case a theoretical discussion is possible only concerning the consequences which such and such a choice of rules would have; and obviously this discussion belongs to the first kind. The special question whether or not a given choice of rules will produce an empiricist language, will then be contained in this set of questions.

In order to make the problem more specific and thereby more simple, let us suppose that we wish to construct L as a physical language, though not as a language for all science. The problems connected with specifically biological or psychological terms, though interesting in themselves, would complicate our present discussion unnecessarily. But the main points of the philosophical discussions of meaning and testability already occur in this specialized case.

In order to formulate the rules of an intended language L, it is necessary to use a language L' which is already available. L' must be given at least practically and need not be stated explicitly as a language-system, that is by formulated rules. We may take as L' the English language. In constructing L, L' serves for two different purposes. First, L' is the syntax-language[23] in which the rules of the object-language L are to be formulated. Secondly, L' may be used as a basis for comparison for L; that is, as a first object-language with which we compare the second object-language L, as to richness of expressions, structure and the like. Thus we may consider the question, to which sentences of the English language (L') do we wish to construct corresponding sentences in L, and to which not. For example, in constructing the language of Principia Mathematica, Whitehead and Russell wished to have available translations for the English sentences of the form "There is something which has the property φ"; they therefore constructed their language-system so as to contain the sentence-form "$(\exists x \cdot \varphi x)$". A difficulty occurs because the English language is not a language-system in the strict sense (i.e. a system of fixed rules) so that the concept of translation cannot be used here in its exact syntactical sense. Nevertheless this concept is sufficiently clear for our present practical purpose. The comparison of L with L' belongs to the rather vague, preliminary considerations which lead to decisions about the system L. Subsequently the result of these decisions can be exactly formulated as rules of the system L.

It is obvious that we are not compelled to construct L so as to contain sentences corresponding to all sentences of L'. If, for example, we wish to construct a language of economics, then its sentences correspond only to a small part of the sentences of the English language L'. But even if L were to be a language adequate for all science there would be many—and I among them—who would not wish to have in

<hr />

[23] Comp. Carnap [4] §1; [5], p. 39.

L a sentence corresponding to every sentence which usually is considered as a correct English sentence and is used by learned people. We should not wish, for example, to have corresponding sentences to many or perhaps most of the sentences occurring in the books of metaphysicians. Or, to give a non-metaphysical example, the members of our Circle did not wish in former times to include into our scientific language a sentence corresponding to the English sentence

S_1: "This stone is now thinking about Vienna."

But at present I should prefer to construct the scientific language in such a way that it contains a sentence S_2 corresponding to S_1. (Of course I should then take S_2 as false, and hence $\sim S_2$ as true.) I do not say that our former view was wrong. Our mistake was simply that we did not recognize the question as one of decision concerning the form of the language; we therefore expressed our view in the form of an assertion—as is customary among philosophers—rather than in the form of a proposal. We used to say: "S_1 is not false but meaningless"; but the careless use of the word 'meaningless' has its dangers and is the second point in which we would like at present to modify the previous formulation.

27. The Principle of Empiricism

It seems to me that it is preferable to formulate the principle of empiricism not in the form of an assertion—"all knowledge is empirical" or "all synthetic sentences that we can know are based on (or connected with) experiences" or the like—but rather in the form of a proposal or requirement. As empiricists, we require the language of science to be restricted in a certain way; we require that descriptive predicates and hence synthetic sentences are not to be admitted unless they have some connection with possible observations, a connection which has to be characterized in a suitable way. By such a formulation, it seems to me, greater clarity will be gained both for carrying on discussion between empiricists and anti-empiricists as well as for the reflections of empiricists.

BIBLIOGRAPHY

For the sake of shortness, the following *publications* will be quoted by the here given figures in square brackets.
* Appeared after the writing of this essay.

AYER,* [1] Language, Truth and Logic, London 1936.

BRIDGMAN, P. W. [1] The Logic of Modern Physics. New York 1927.

CARNAP, R. [1] Der logische Aufbau der Welt. Berlin (now F. Meiner, Leipzig) 1928.

[2a] Die physikalische Sprache als Universalsprache der Wissenschaft. Erkenntnis 2, 1932.

[2b] (Translation:) The Unity of Science. Kegan Paul, London 1934.

[3] Ueber Protokollsätze. Erkenntnis 3, 1932.

[4a] Logische Syntax der Sprache. Springer, Wien 1934.

[4b] (Translation:) Logical Syntax of Language. Kegan Paul, London, (Harcourt, Brace, New York) 1936.

[5] Philosophy and Logical Syntax. Kegan Paul, London 1935.

[6] Formalwissenschaft und Realwissenschaft. Erkenntnis 5, 1935. (*Congress* [1]).

[7] Ueber ein Gültigkeitskriterium für die Sätze der klassischen Mathematik. Monatsh. Math. Phys. 42, 1935.

[8] Les Concepts Psychologiques et les Concepts Physiques sont-ils Foncièrement Différents? Revue de Synthèse 10, 1935.

[9] Wahrheit und Bewährung. In: *Congress* [3].

[10] Von der Erkenntnistheorie zur Wissenschaftslogik. In: *Congress* [3].

[11] Ueber die Einheitssprache der Wissenschaft. Logische Bemerkungen zur Enzyklopädie. In: *Congress* [3].

[12] Gibt es nicht-prüfbare Voraussetzungen der Wissenschaft? Scientia, 1936.

CONGRESS [1] Einheit der Wissenschaft. Bericht über die Prager Vorkonferenz der Internationalen Kongresse für Einheit der Wissenschaft, Sept. 1934. Erkenntnis 5, Heft 1–3, 1935.

*[2] Erster Internationaler Kongress für Einheit der Wissenschaft (Congrès Internat. de Philos. Scientifique), Paris 1935. [Report of Sessions.] Erkenntnis 5, Heft 6, 1936.

*[3] Actes du Ier Congrès Internat. de Philos. Scientifique, Paris 1935. 8 fasc. Hermann & Cie, Paris 1936.

DUCASSE, C. J. *[1] Verification, Verifiability and Meaningfulness. Journ. of Philos. 33, 1936.

FEIGL, H. *[1] Sense and Nonsense in Scientific Realism. In: *Congress* [3].

FRANK, P. H. [1] Das Kausalgesetz und seine Grenzen. Springer, Wien, 1932.

HEMPEL, C. G. [1] Beiträge zur logischen Analyse des Wahrheitsbegriffs. Diss. Berlin, 1934.

[2] Ueber den Gehalt von Wahrscheinlichkeitsaussagen. Erkenntnis 5, 1935.

[3] On the Logical Positivist's Theory of Truth. Analysis 2, 1935.

[4] Some Remarks on 'Facts' and Propositions. Analysis 2, 1935.

HILBERT, D. und ACKERMANN, W. [1] Grundzüge der theoretischen Logik. Springer, Berlin, 1928.

KAUFMANN, F. [1] Das Unendliche in der Mathematik und seine Ausschaltung. Deuticke, Wien 1930.

LEWIS, C. I. [1] with LANGFORD, C. H. Symbolic Logic. The Century Co., New York 1932.
[2] Experience and Meaning. Philos. Review 43, 1934.

MORRIS, CH. W. [1] Philosophy of Science and Science of Philosophy. Philos. of Sc. 2, 1935.
[2] The Concept of Meaning in Pragmatism and Logical Positivism. Proc. 8th Internat. Congr. Philos. (1934). Prague 1936.
[3] Semiotic and Scientific Empiricism. In: *Congress* [3].

NAGEL, E. [1] Verifiability, Truth, and Verification. Journ. of Philos. 31, 1934.
*[2] Impressions and Appraisals of Analytic Philosophy in Europe. Journ. of Philos. 33, 1936.

NESS, A. *[1] Erkenntnis und wissenschaftliches Verhalten. Norske Vid.-Akad. II. Hist.-Fil. Kl., No. 1. Oslo 1936.

NEURATH, O. [1] Physicalism. Monist 41, 1931.
[2] Physikalismus. Scientia 50, 1931.
[3] Soziologie im Physikalismus. Erkenntnis 2, 1931.
[4] Protokollsätze. Erkenntnis 3, 1932.
[5] Radikaler Physikalismus und "wirkliche Welt." Erkenntnis 4, 1934.
[6] Pseudorationalismus der Falsifikation. Erkenntnis 5, 1935.
*[7] Le Développement du Cercle du Vienne et l'Avenir de l'Empirisme Logique. Hermann, Paris 1935.
*[8] Einzelwissenschaften, Einheitswissenschaft, Pseudorationalismus. In: *Congress* [3].

POPPER, K. [1] Logik der Forschung. Springer, Wien 1935.
*[2] Empirische Methode. In: *Congress* [3].

RAMSEY, F. P. [1] General Propositions and Causality. 1929. Published posthumously in: The Foundations of Mathematics, and other Logical Essays, p. 237–255. Harcourt, Brace, New York 1931.

REICHENBACH, H. [1] Wahrscheinlichkeitslehre. Sijthoff, Leyden 1935.
*[2] Ueber Induktion und Wahrscheinlichkeit. Erkenntnis 5, 1935.
*[3] Logistic Empiricism in Germany and the Present State of its Problems. Journ. of Philos. 33, 1936.
*[4] L'Empirisme Logistique et la Désaggrégation de l'Apriori. In: *Congress* [3].

RUSSELL, B. [1] see *Whitehead*.

[2] Our Knowledge of the External World. Open Court, Chicago and London 1914.

RUSSELL, L. J. [1] Communication and Verification. Proc. Arist. Soc., Suppl. Vol. 13, 1934.

SCHLICK, M. [1] Die Kausalität in der gegenwärtigen Physik. Naturwiss. 19, 1931.
[2] Ueber das Fundament der Erkenntnis. Erkenntnis 4, 1934.
[3] Facts and Propositions. Analysis 2, 1935.
[4] Meaning and Verification. Philos. Review 45, 1936.

STACE, W. T. *[1] Metaphysics and Meaning. Mind 44, 1935.

STEBBING, S. L. [1] Communication and Verification. Proc. Arist. Soc., Suppl. Vol. 13, 1934.

TARSKI, A. *[1] Der Wahrheitsbegriff in den formalisierten Sprachen. Stud. Philos. 1, 1936.

WAISMANN, F. [1] Logische Analyse des Wahrscheinlichkeitsbegriffs. Erkenntnis 1, 1930.

WEYL, H. [1] Die heutige Erkenntnislage in der Mathematik. Symposion 1, 1925; also published separately.

WHITEHEAD, A. N. and RUSSELL, B. [1] Principia Mathematica. (1910–12) 2nd ed. Cambridge 1925–27.

WITTGENSTEIN, L. [1] Tractatus Logico-Philosophicus. Harcourt, Brace, New York 1922.

R. B. BRAITHWAITE

The Nature of Theoretical Concepts and the Role of Models in an Advanced Science

The function of every science is to establish laws—true hypotheses—which cover the behaviour of the observable things and events which are the subject-matter of the science thereby enabling the scientist both to connect together his knowledge of particular events and to enable him to predict what events will happen under certain circumstances. If these circumstances can be produced at will, knowledge of the general laws will enable to some extent the applied scientist to control the course of nature and to construct machines or other artifacts which will behave in known ways.

An advanced science like physics is not content only with establishing lowest-level generalisations covering physical events: it aims at, and has been largely successful in, subsuming its lowest-level generalisations under higher-level hypotheses, and thus in organising its hypotheses into a hierarchical deductive system—a scientific theory—in which a hypothesis at a lower level is shown to be deducible from a set of hypotheses at a higher level. Notable examples of this are Maxwell's subsumption of the laws of optics under his very general electromagnetic equations and Einstein's subsumption of gravitational laws under his General Theory of Relativity.

The concepts which enter into the higher-level hypotheses of an advanced science are usually concepts (e.g. electric-field vector, electron, Schrödinger wave-function) which are not directly observable things or properties as are those which appear in the lowest-level generalisations of the science: instead they are theoretical concepts which appear at the beginning of the deductive theory but which are eliminated in the course of the deduction. These theoretical concepts present a prob-

* Reprinted by kind permission of the author and publisher from the *Revue Internationale de Philosophie*, Vol. 8 (1954), 34–40.

lem to the philosopher of science: namely, what is their epistemological status? They are clearly in some way empirical concepts—an electron or a Schrödinger wave-function is not an object of pure mathematics like a prime number; but an electron is not observable in the sense in which a flash of light or the pointer reading on a measuring scale is observable. Nevertheless the truth of propositions about electrons is tested by the observable behaviour of measuring instruments; and the question is therefore in what manner an electron or other theoretical concept is an empirical concept.

An answer to this question, an answer implicit in the writings of many philosophers of science such as Ernst Mach and Karl Pearson, was given explicitly by Bertrand Russell in his doctrine of 'logical constructions'. "The supreme maxim in scientific philosophising is this: Wherever possible, logical constructions are to be substituted for inferred entities." (*Mysticism and Logic and other essays*, 1918, p. 155.) According to the logical construction view electrons, for example, are logical constructions out of the observed events and objects by which their presence can be detected; this is equivalent to saying that the word "electron" can be explicitly defined in terms of such observations. On this view every sentence containing the word "electron" is translatable, without loss of meaning, into a sentence in which there only occur words which denote entities (events, objects, properties, relations) which are directly observable. It is the business of a philosopher of science to show how these translations are to be made, and thus to show how the theoretical terms of a science can be explicitly defined by means of observable entities. A philosopher of physics should be able to make this translation in the case of the word "electron", and thus be able to exhibit the way in which electrons are logical constructions out of observable entities. Russell's programme of logical construction is similar to the 'operationalist' programme proposed by P. W. Bridgman, according to which theoretical terms must be defined by means of the empirical 'operations' involved in their measurement (*The Logic of Modern Physics*, 1927).

This 'logical construction' view of the nature of theoretical concepts was criticised by F. P. Ramsey in some notes which he wrote in 1929, a few months before his death at the age of 26, and which were published posthumously (*The Foundations of Mathematics and other logical essays*, 1931, pp. 212 ff.). Ramsey developed his criticism by constructing a simple example of a scientific theory; I have been able to construct (in my *Scientific Explanation*, 1953, Chapter III) even simpler examples which show precisely the defects of the logical construction view.

The point displayed by both Ramsey's and my examples is that, although it is always possible to define the theoretical terms occurring in the highest-level hypotheses of a theory by means of the terms denoting directly observable entities which occur in the lowest-level generalisations which the theory was propounded to explain, such a definition will prevent the theory from being expanded into a wider theory capable of explaining new lowest-level generalisations which may subsequently be established. To treat theoretical concepts as logical constructions out of observable entities would be to *ossify* the scientific theory in which they occur: the theory would be adequate to cover systematically a particular set of lowest-level generalisations already established, but there would be no hope of extending the theory to explain more generalisations than it was originally designed to explain. A scientific theory to be capable (like all good scientific theories) of this sort of growth must give more freedom of play to its theoretical concepts than the logical construction view will allow them to have.

What then, if the logical construction view is inadequate, is the epistemological status of theoretical concepts? A way of answering this question, essentially that of Ramsey, is to say that the status of the theoretical concept *electron* is given by specifying its place in the deductive system of contemporary physics in the following way: there is a property (called "being an electron") which is such that from certain higher-level hypotheses which are true and which are about the property *electron* there follow certain lowest-level generalisations which are empirically testable. According to this answer, nothing is asserted about the 'nature' of the property E in itself; all that is asserted is that there are instances of E, namely electrons. To say that electrons exist is to assert the truth of the physical theory in which there occurs the concept of *being an electron*.

There is, however, another way of answering the status-question which is open to a philosopher of science who, with knowledge of the work of logicians such as Carnap done since Ramsey's death, would wish to make a sharper distinction than was made by Ramsey between a scientific theory arranged as a *deductive system* and the *calculus* (or language) representing the deductive system. So let us consider, not the nature of the theoretical concept *electron* in the deductively-arranged physical theory, but the role played by the term *electron* (or other synonymous symbols) in the calculus representing the theory. This calculus consists of a series of formulae (or sentences) arranged in such a way that all the formulae, except for a small number of formulae (called "initial formulae"), are derived from these initial formulae in accordance with the rules of the calculus. The calculus will be interpreted to represent the physical deductive theory by taking the highest-level hypotheses of the theory to be represented by initial formulae, and lower-level hypotheses and generalisations to be represented by derived formulae in the calculus. For this interpretation to be possible it is necessary that the rules of the calculus should correspond to the logical and mathematical principles of deduction used in making the deductions within the deductive theory.

When the deductive system to be represented by the calculus is a pure one, that is a system containing only logically necessary propositions (e.g. a deductive system of arithmetic starting with Peano's axioms), the calculus is interpreted all in a piece. Meanings are attached to the 'primitive' symbols occurring in the initial formulae representing the axioms (e.g. to "number", "successor of", "zero" in Peano's arithmetical calculus), and meanings are attached to all the other symbols used in the calculus by the use of formulae of the calculus interpreted as explicit definitions of these other symbols in terms of the primitive symbols. (These definitions may, of course, be 'contextual definitions,' that is, definitions of a whole formula rather than of a separate term.) The meanings attached to the symbols do not depend upon the order of the formulae in the calculus representing the pure deductive system: the meaning, for example, of "prime number" does not depend upon the order in which theorems about prime numbers are deduced from the axioms of arithmetic.

The situation is quite different in the case in which the deductive system to be represented by the calculus is a scientific theory containing empirical hypotheses at different levels. Here, on my view, the calculus is not interpreted all in a piece: meanings are first attached to the symbols denoting the directly observable properties and relations which occur in the directly testable lowest-level generalisations of the theory. Meaning is then attached to the symbols which are to denote the theoretical concepts of the theory merely by virtue of the fact that they occur in the initial formulae of the calculus representing the theory. The formulae of the calculus are arranged in an order corresponding to the deductive arrangement of the hypotheses of the theory, with the initial formulae corresponding to the highest-level hypotheses. The initial formulae are interpreted as representing propositions from which directly testable propositions logically follow, and the symbols (e.g. *electrons*) which occur in these initial formulae are interpreted as being essential parts of these formulae. No direct meaning is attached to the term *electron:* it is given a meaning indirectly by the function it plays in the calculus which is interpreted as representing the physical theory.

On this view the status of the concept *electron,* and the question "Do electrons really exist?" can only be discussed in terms of the role played by the word *electron* in the exposition of physical theory. This view resembles Russell's 'logical construction' view in that in both cases what is in question is the meaning of the word or other symbol. But whereas Russell would say that the word is given a meaning by being *explicitly* defined by means of words denoting observable entities, I would say that it is given a meaning by showing its place in a calculus representing a scientific theory. This may be called giving an *implicit* or a *contextual* definition of the theoretical terms by means of words denoting observable entities occurring in the formulae of the calculus which represent the directly testable lowest-level generalisations of the theory; and my account may be taken as an elucidation of such indirect or contextual definition. (It will give a wider sense of "contextual definition" than the usual one, for it is the whole interpreted calculus and not only one sentence [or type of sentence] which will form the 'context' for the purpose of the contextual definition.)

A direction of attention upon the calculus representing a deductive scientific theory will also throw light upon the use of a *model* in thinking about the scientific theory. Suppose that the calculus which is interpreted as representing the theory can also be interpreted as a deductive system in such a way that a direct meaning is given to all the symbols occurring in its initial formulae. In this second interpretation of the calculus the propositions represented will contain no 'theoretical' concepts, and the calculus will be interpreted all in a piece —as in the case of a calculus representing a pure mathematical, deductive system. Thus this interpretation does not present the epistemological difficulty presented by the original interpretation of the calculus as representing the scientific theory. The deductive system which is this second interpretation of the calculus may be regarded as a model for the theory. A theory and a model for it have the same formal structure, since they are both represented by one and the same calculus. There is a one-one correlation between the propositions of the theory and those of the model, and the deductive arrangement of the propositions in the theory corresponds to that of the correlated propositions in the model (and vice versa). But the epistemological structure of the theory and of a model for it are different: in order to give the model the calculus is directly interpreted all in a piece, whereas to give the theory it is derived formulae of the calculus which are directly interpreted, the earlier formulae being interpreted indirectly by virtue of their place in the calculus.

The fact that a model has the same logical structure as, but a simpler epistemological structure than, the theory for which it is a model explains the use of models in thinking about a theory in an advanced science. For thinking about the model will, for many purposes, serve as a substitute for thinking explicitly about the calculus of which both theory and model are interpretations, since the model is a quite straightforward interpretation of the calculus. So to think of the model instead of the calculus in connection with the theory avoids the self-consciousness required in thinking at the same time of a theory and of the language in which it is expounded, and thus allows of a philosophically unsophisticated approach to an understanding of the logical structure of a scientific deductive theory.

The dangers of thinking of a theory by way of thinking of a model for it are, first, that we may, if we are not careful, suppose that there are concepts involved in the theory which correspond to *all* the properties of the objects in the model; we may think, for example, that the electrons in an atom have all the spatial properties of the balls in a 'solar system' model of an atom. A second—more subtle—danger is that it may well happen that some of the propositions in the model which are the interpretations of the initial formulae of the calculus are logically necessary propositions (indeed they may all be logically necessary propositions, in which case the model is a pure mathematical model). We may then be tempted illicitly to transfer the logical necessity of these propositions in the model on to the correlated propositions in the theory, and thus to suppose that some or all of the highest-level hypotheses of the theory are logically necessary instead of contingent. In using models we must never forget that we are engaging in *as-if* thinking: the theoretical concepts in a scientific theory behave as if they were elements in the model, but only in certain respects. To forget the limitations is to misuse the valuable aid to thought provided by the model.

The topics of this paper are discussed at greater length in Chapters III and IV of my book *Scientific Explanation* (Cambridge: at the University Press: 1953).

King's College, Cambridge

C. G. HEMPEL

Empiricist Criteria of Cognitive Significance: Problems and Changes

1. THE GENERAL EMPIRICIST CONCEPTION OF COGNITIVE AND EMPIRICAL SIGNIFICANCE

It is a basic principle of contemporary empiricism that a sentence makes a cognitively significant assertion, and thus can be said to be either true or false, if and only if either (1) it is analytic or contradictory—in which case it is said to have purely logical meaning or significance—or else (2) it is capable, at least potentially, of test by experiential evidence—in which case it is said to have empirical meaning or significance. The basic tenet of this principle, and especially of its second part, the so-called testability criterion of empirical meaning (or better: meaningfulness), is not peculiar to empiricism alone: it is characteristic also of contemporary operationism, and in a sense of pragmatism as well; for the pragmatist maxim that a difference must make a difference to be a difference may well be construed as insisting that a verbal difference between two sentences must make a difference in experiential implications if it is to reflect a difference in meaning.

How this general conception of cognitively significant discourse led to the rejection, as devoid of logical and empirical meaning, of various formulations in speculative metaphysics, and even of certain hypoth-

This essay combines, with certain omissions and some other changes, the contents of two articles: "Problems and Changes in the Empiricist Criterion of Meaning," *Revue Internationale de Philosophie* No. 11, pp. 41–63 (January, 1950); and "The Concept of Cognitive Significance: A Reconsideration," *Proceedings of the American Academy of Arts and Sciences* 80, No. 1, pp. 61–77 (1951). This material is reprinted with kind permission of the Director of *Revue Internationale de Philosophie* and of the American Academy of Arts and Sciences.

eses offered within empirical science, is too well known to require recounting. I think that the general intent of the empiricist criterion of meaning is basically sound, and that notwithstanding much over-simplification in its use, its critical application has been, on the whole, enlightening and salutary. I feel less confident, however, about the possibility of restating the general idea in the form of precise and general criteria which establish sharp dividing lines (a) between statements of purely logical and statements of empirical significance, and (b) between those sentences which do have cognitive significance and those which do not.

In the present paper, I propose to reconsider these distinctions as conceived in recent empiricism, and to point out some of the difficulties they present. The discussion will concern mainly the second of the two distinctions; in regard to the first, I shall limit myself to a few brief remarks.

2. THE EARLIER TESTABILITY CRITERIA OF MEANING AND THEIR SHORTCOMINGS

Let us note first that any general criterion of cognitive significance will have to meet certain requirements if it is to be at all acceptable. Of these, we note one, which we shall consider here as expressing a necessary, though by no means sufficient, *condition of adequacy* for criteria of cognitive significance.

(A) If under a given criterion of cognitive significance, a sentence N is nonsignificant, then so must be all truth-functional compound sentences in which N occurs nonvacuously as a component.* For if N cannot be significantly assigned a truth value, then it is impossible to assign truth values to the compound sentences containing N; hence, they should be qualified as nonsignificant as well.

We note two corollaries of requirement (A):

(A1) If under a given criterion of cognitive significance, a sentence S is nonsignificant, then so must be its negation, $\sim S$.

(A2) If under a given criterion of cognitive significance, a sentence N is nonsignificant, then so must be any conjunction $N \cdot S$ and any disjunction $N \vee S$, no matter whether S is significant under the given criterion or not.

We now turn to the initial attempts made in recent empiricism to establish general criteria of cognitive significance. Those attempts were governed by the consideration that a sentence, to make an empirical

---

* Truth-functional compound sentences in which N occurs are those sentences formed by use of the connectives 'and', 'or', 'if–then', 'if and only if' and 'not'. N occurs nonvacuously in a sentence S if the truth value of S depends (partially) on the truth value of N. For example, N occurs nonvacuously in $N \vee M$, but occurs vacuously in $M v \sim M \vee N$, since the sentence is true whether or not N is. —Ed.

assertion must be capable of being borne out by, or conflicting with, phenomena which are potentially capable of being directly observed. Sentences describing such potentially observable phenomena—no matter whether the latter do actually occur or not—may be called observation sentences. More specifically, an *observation sentence* might be construed as a sentence—no matter whether true or false—which asserts or denies that a specified object, or group of objects, of macroscopic size has a particular *observable characteristic,* that is, a characteristic whose presence or absence can, under favorable circumstances, be ascertained by direct observation.[1]

The task of setting up criteria of empirical significance is thus transformed into the problem of characterizing in a precise manner the relationship which obtains between a hypothesis and one or more observation sentences whenever the phenomena described by the latter either confirm or disconfirm the hypothesis in question. The ability of a given sentence to enter into that relationship to some set of observation sentences would then characterize its testability-in-principle, and thus its empirical significance. Let us now briefly examine the major attempts that have been made to obtain criteria of significance in this manner.

One of the earliest criteria is expressed in the so-called *verifiability requirement.* According to it, a sentence is empirically significant if and only if it is not analytic and is capable, at least in principle, of complete verification by observational evidence; i.e., if observational

---

[1] Observation sentences of this kind belong to what Carnap has called the thing-language, cf., e.g. (1938), pp. 52–53. That they are adequate to formulate the data which serve as the basis for empirical tests is clear in particular for the intersubjective testing procedures used in science as well as in large areas of empirical inquiry on the common-sense level. In epistemological discussions, it is frequently assumed that the ultimate evidence for beliefs about empirical matters consists in perceptions and sensations whose description calls for a phenomenalistic type of language. The specific problems connected with the phenomenalistic approach cannot be discussed here; but it should be mentioned that at any rate all the critical considerations presented in this article in regard to the testability criterion are applicable, *mutatis mutandis,* to the case of a phenomenalistic basis as well.

[2] Originally, the permissible evidence was meant to be restricted to what is observable by the speaker and perhaps his fellow beings during their life times. Thus construed, the criterion rules out, as cognitively meaningless, all statements about the distant future or the remote past, as has been pointed out, among others, by Ayer (1946), Chapter I; by Pap (1949), Chapter 13, esp. pp. 333 ff.; and by Russell (1948), pp. 445–47. This difficulty is avoided, however, if we permit the evidence to consist of any finite set of "logically possible observation data", each of them formulated in an observation sentence. Thus, e.g., the sentence S_1, "The tongue of the largest dinosaur in New York's Museum of Natural History was blue or black" is completely verifiable in our sense; for it is a logical consequence of the sentence S_2, "The tongue of the largest dinosaur in New York's Museum of Natural History was blue"; and this is an observation sentence, in the sense just indicated.

evidence can be described which, if actually obtained, would conclusively establish the truth of the sentence.[2] With the help of the concept of observation sentence, we can restate this requirement as follows: A sentence S has empirical meaning if and only if it is possible to indicate a finite set of observation sentences, $O_1, O_2, \ldots, O_n$, such that if these are true, then S is necessarily true, too. As stated, however, this condition is satisfied also if S is an analytic sentence or if the given observation sentences are logically incompatible with each other. By the following formulation, we rule these cases out and at the same time express the intended criterion more precisely:

(2.1) *Requirement of Complete Verifiability in Principle.* A sentence has empirical meaning if and only if it is not analytic and follows

---

And if the concept of *verifiability in principle* and the more general concept of *confirmability in principle*, which will be considered later, are construed as referring to *logically possible evidence* as expressed by observation sentences, then it follows similarly that the class of statements which are verifiable, or at least confirmable, in principle include such assertions as that the planet Neptune and the Antarctic Continent existed before they were discovered, and that atomic warfare, if not checked, will lead to the extermination of this planet. The objections which Russell (1948), pp. 445 and 447, raises against the verifiability criterion by reference to those examples do not apply therefore if the criterion is understood in the manner here suggested. Incidentally, statements of the kind mentioned by Russell, which are not actually verifiable by any human being, were explicitly recognized as cognitively significant already by Schlick (1936), Part V, who argued that the impossibility of verifying them was "merely empirical." The characterization of verifiability with the help of the concept of observation sentence as suggested here might serve as a more explicit and rigorous statement of that conception.

[3] As has frequently been emphasized in the empiricist literature, the term "verifiability" is to indicate, of course, the conceivability, or better, the logical possibility, of evidence of an observational kind which, if actually encountered, would constitute conclusive evidence for the given sentence; it is not intended to mean the technical possibility of performing the tests needed to obtain such evidence, and even less the possibility of actually finding directly observable phenomena which constitute conclusive evidence for that sentence—which would be tantamount to the actual existence of such evidence and would thus imply the truth of the given sentence. Analogous remarks apply to the terms "falsifiability" and "confirmability". This point has clearly been disregarded in some critical discussions of the verifiability criterion. Thus, e.g., Russell (1948), p. 448 construes verifiability as the actual existence of a set of conclusively verifying occurrences. This conception, which has never been advocated by any logical empiricist, must naturally turn out to be inadequate since according to it the empirical meaningfulness of a sentence could not be established without gathering empirical evidence, and moreover enough of it to permit a conclusive proof of the sentence in question! It is not surprising, therefore, that his extraordinary interpretation of verifiability leads Russell to the conclusion: "In fact, that a proposition is verifiable is itself not verifiable" (*l.c.*). Actually, under the empiricist interpretation of complete verifiability, any statement asserting the verifiability of some sentence S whose text is quoted, is either analytic or contradictory; for the decision whether there exists a class of observation sentences which entail S, i.e., whether such observation sentences can be formulated, no matter whether they are true or false—that decision is a purely logical matter.

logically from some finite and logically consistent class of observation sentences.[3] These observation sentences need not be true, for what the criterion is to explicate is testability by "potentially observable phenomena," or testability "in principle".

In accordance with the general conception of cognitive significance outlined earlier, a sentence will now be classified as cognitively significant if either it is analytic or contradictory, or it satisfies the verifiability requirement.

This criterion, however, has several serious defects. One of them has been noted by several writers:

a. Let us assume that the properties of being a stork and of being red-legged are both observable characteristics, and that the former does not logically entail the latter. Then the sentence

(S1)   All storks are red-legged

is neither analytic nor contradictory; and clearly, it is not deducible from a finite set of observation sentences. Hence, under the contemplated criterion, S1 is devoid of empirical significance; and so are all other sentences purporting to express universal regularities or general laws. And since sentences of this type constitute an integral part of scientific theories, the verifiability requirement must be regarded as overly restrictive in this respect.

Similarly, the criterion disqualifies all sentences such as 'For any substance there exists some solvent', which contain both universal and existential quantifiers (i.e., occurrences of the terms 'all' and 'some' or their equivalents); for no sentences of this kind can be logically deduced from any finite set of observation sentences.

Two further defects of the verifiability requirement do not seem to have been widely noticed:

b. As is readily seen, the negation of S1

(∼S1)   There exists at least one stork that is not red-legged

is deducible from any two observation sentences of the type 'a is a stork' and 'a is not red-legged'. Hence, ∼S1 is cognitively significant under our criterion, but S1 is not, and this constitutes a violation of condition (A1).

c. Let S be a sentence which does, and N a sentence which does not satisfy the verifiability requirement. Then S is deducible from some set of observation sentences; hence, by a familiar rule of logic, SvN is

---

[4] The arguments here adduced against the verifiability criterion also prove the inadequacy of a view closely related to it, namely that two sentences have the same cognitive significance if any set of observation sentences which would verify one of them would also verify the other, and conversely. Thus, e.g., under this criterion, any two general laws would have to be assigned the same cognitive significance, for no general law is verified by any set of observation sentences. The view just referred

deducible from the same set, and therefore cognitively significant according to our criterion. This violates condition (A2) above.[4]

Strictly analogous considerations apply to an alternative criterion, which makes complete falsifiability in principle the defining characteristic of empirical significance. Let us formulate this criterion as follows:

(2.2) *Requirement of Complete Falsifiability in Principle.* A sentence has empirical meaning if and only if its negation is not analytic and follows logically from some finite logically consistent class of observation sentences.

This criterion qualifies a sentence as empirically meaningful if its negation satisfies the requirement of complete verifiability; as it is to be expected, it is therefore inadequate on similar grounds as the latter:

(a) It denies cognitive significance to purely existential hypotheses, such as 'There exists at least one unicorn', and all sentences whose formulation calls for mixed—i.e., universal and existential—quantification, such as 'For every compound there exists some solvent', for none of these can possibly be conclusively falsified by a finite number of observation sentences.

(b) If '*P*' is an observation predicate, then the assertion that all things have the property *P* is qualified as significant, but its negation, being equivalent to a purely existential hypothesis, is disqualified [cf. (*a*)]. Hence, criterion (2.2) gives rise to the same dilemma as (2.1).

(c) If a sentence *S* is completely falsifiable whereas *N* is a sentence which is not, then their conjunction, *S·N* (i.e., the expression obtained by connecting the two sentences by the word 'and') is completely falsifiable; for if the negation of *S* is entailed by a class of observation sentences, then the negation of *S·N* is, *a fortiori*, entailed by the same class. Thus, the criterion allows empirical significance to many sentences which an adequate empiricist criterion should rule out, such as 'All swans are white and the absolute is perfect.'

In sum, then, interpretations of the testability criterion in terms of complete verifiability or of complete falsifiability are inadequate because they are overly restrictive in one direction and overly inclusive in another, and because both of them violate the fundamental requirement A.

Several attempts have been made to avoid these difficulties by con-

to must be clearly distinguished from a position which Russell examines in his critical discussion of the positivistic meaning criterion. It is "the theory that two propositions whose verified consequences are identical have the same significance" (1948), p. 448. This view is untenable indeed, for what consequences of a statement have actually been verified at a given time is obviously a matter of historical accident which cannot possibly serve to establish identity of cognitive significance. But I am not aware that any logical empiricist ever subscribed to that "theory."

struing the testability criterion as demanding merely a partial and possibly indirect confirmability of empirical hypotheses by observational evidence.

A formulation suggested by Ayer[5] is characteristic of these attempts to set up a clear and sufficiently comprehensive criterion of confirmability. It states, in effect, that a sentence *S* has empirical import if from *S* in conjunction with suitable subsidiary hypotheses it is possible to derive observation sentences which are not derivable from the subsidiary hypotheses alone.

This condition is suggested by a closer consideration of the logical structure of scientific testing; but it is much too liberal as it stands. Indeed, as Ayer himself has pointed out in the second edition of his book, *Language, Truth, and Logic*,[6] his criterion allows empirical import to any sentence whatever. Thus, for example, if *S* is the sentence 'The absolute is perfect', it suffices to choose as a subsidiary hypothesis the sentence 'If the absolute is perfect then this apple is red' in order to make possible the deduction of the observation sentence 'This apple is red', which clearly does not follow from the subsidiary hypothesis alone.

To meet this objection, Ayer proposed a modified version of his testability criterion. In effect, the modification restricts the subsidiary hypotheses mentioned in the previous version to sentences which either are analytic or can independently be shown to be testable in the sense of the modified criterion.[7]

But it can readily be shown that this new criterion, like the requirement of complete falsifiability, allows empirical significance to any conjunction $S \cdot N$, where *S* satisfies Ayer's criterion while *N* is a sentence such as 'The absolute is perfect', which is to be disqualified by that criterion. Indeed, whatever consequences can be deduced from *S* with the help of permissible subsidiary hypotheses can also be deduced from $S \cdot N$ by means of the same subsidiary hypotheses; and as Ayer's new criterion is formulated essentially in terms of the deducibility of a certain type of consequence from the given sentence, it countenances $S \cdot N$ together with *S*. Another difficulty has been pointed

---

[5] (1936, 1946), Chap. I. The case against the requirements of verifiability and of falsifiability, and in favor of a requirement of partial confirmability and disconfirmability, is very clearly presented also by Pap (1949), chapter 13.

[6] (1946), 2d ed., pp. 11–12.

[7] This restriction is expressed in recursive form and involves no vicious circle. For the full statement of Ayer's criterion, see Ayer (1946), p. 13.

[8] Church (1949). An alternative criterion recently suggested by O'Connor (1950) as a revision of Ayer's formulation is subject to a slight variant of Church's stricture: It can be shown that if there are three observation sentences none of which entails any of the others, and if *S* is any noncompound sentence, then either *S* or $\backsim S$ is significant under O'Connor's criterion.

out by Church, who has shown[8] that if there are any three observation sentences none of which alone entails any of the others, then it follows for any sentence S whatsoever that either it or its denial has empirical import according to Ayer's revised criterion.

All the criteria considered so far attempt to explicate the concept of empirical significance by specifying certain logical connections which must obtain between a significant sentence and suitable observation sentences. It seems now that this type of approach offers little hope for the attainment of precise criteria of meaningfulness: this conclusion is suggested by the preceding survey of some representative attempts, and it receives additional support from certain further considerations, some of which will be presented in the following sections.

3. CHARACTERIZATION OF SIGNIFICANT SENTENCES BY CRITERIA FOR THEIR CONSTITUENT TERMS

An alternative procedure suggests itself which again seems to reflect well the general viewpoint of empiricism: It might be possible to characterize cognitively significant sentences by certain conditions which their constituent terms have to satisfy. Specifically, it would seem reasonable to say that all extralogical terms[9] in a significant sentence must have experiential reference, and that therefore their meanings must be capable of explication by reference to observables exclusively.[10] In order to exhibit certain analogies between this approach and the previous one, we adopt the following terminological conventions:

Any term that may occur in a cognitively significant sentence will be called a *cognitively significant term*. Furthermore, we shall understand by an *observation term* any term which either (a) is an *observation predicate,* that is, signifies some observable characteristic (as do the terms 'blue', 'warm', 'soft', 'coincident with', 'of greater apparent brightness than') or (b) names some physical object of macroscopic size (as do the terms 'the needle of this instrument', 'the Moon', 'Krakatoa Volcano', 'Greenwich, England', 'Julius Caesar').

Now while the testability criteria of meaning aimed at characterizing the cognitively significant sentences by means of certain inferential connections in which they must stand to some observation sen-

---

[9] An extralogical term is one that does not belong to the specific vocabulary of logic. The following phrases, and those definable by means of them, are typical examples of logical terms: 'not', 'or', 'if . . . then', 'all', 'some', '. . . is an element of class . . .' . Whether it is possible to make a sharp theoretical distinction between logical and extralogical terms is a controversial issue related to the problem of discriminating between analytic and synthetic sentences. For the purpose at hand, we may simply assume that the logical vocabulary is given by enumeration.

[10] For a detailed exposition and critical discussion of this idea, see H. Feigl's stimulating and enlightening article (1950).

tences, the alternative approach under consideration would instead try to specify the vocabulary that may be used in forming significant sentences. This vocabulary, the class of significant terms, would be characterized by the condition that each of its elements is either a logical term or else a term with empirical significance; in the latter case, it has to stand in certain definitional or explicative connections to some observation terms. This approach certainly avoids any violations of our earlier conditions of adequacy. Thus, for example, if S is a significant sentence, that is, contains cognitively significant terms only, then so is its denial, since the denial sign, and its verbal equivalents, belong to the vocabulary of logic and are thus significant. Again, if N is a sentence containing a nonsignificant term, then so is any compound sentence which contains N.

But this is not sufficient, of course. Rather, we shall now have to consider a crucial question analogous to that raised by the previous approach: Precisely how are the logical connections between empirically significant terms and observation terms to be construed if an adequate criterion of cognitive significance is to result? Let us consider some possibilities.

(3.1) The simplest criterion that suggests itself might be called the *requirement of definability*. It would demand that any term with empirical significance must be explicitly definable by means of observation terms.

This criterion would seem to accord well with the maxim of operationism that all significant terms of empirical science must be introduced by operational definitions. However, the requirement of definability is vastly too restrictive, for many important terms of scientific and even pre-scientific discourse cannot be explicitly defined by means of observation terms.

In fact, as Carnap[11] has pointed out, an attempt to provide explicit definitions in terms of observables encounters serious difficulties as soon as disposition terms, such as 'soluble', 'malleable', 'electric conductor', etc., have to be accounted for; and many of these occur even on the pre-scientific level of discourse.

Consider, for example, the word 'fragile'. One might try to define it by saying that an object x is fragile if and only if it satisfies the following condition: If at any time t the object is sharply struck, then it breaks at that time. But if the statement connectives in this phrasing are construed truth-functionally, so that the definition can be symbolized by

$$(D) \qquad Fx \equiv (t)(Sxt \supset Bxt)$$

then the predicate 'F' thus defined does not have the intended mean-

---

[11] Cf. (1936–37), especially section 7. Reprinted in this volume on pp. 27–46.

ing. For let a be any object which is not fragile (e.g., a raindrop or a rubber band), but which happens not to be sharply struck at any time throughout its existence. Then 'Sat' is false and hence '$Sat \supset Bat$' is true for all values of 't'; consequently, 'Fa' is true though a is not fragile.

To remedy this defect, one might construe the phrase 'if . . . then . . .' in the original definiens as having a more restrictive meaning than the truth-functional conditional. This meaning might be suggested by the subjunctive phrasing 'If x were to be sharply struck at any time t, then x would break at t.' But a satisfactory elaboration of this construal would require a clarification of the meaning and the logic of counterfactual and subjunctive conditionals, which is a thorny problem.[12] *

An alternative procedure was suggested by Carnap in his theory of reduction sentences.[13] These are sentences which, unlike definitions, specify the meaning of a term only conditionally or partially. The term 'fragile', for example, might be introduced by the following reduction sentence:

$$(R) \qquad\qquad (x)\,(t)\,[Sxt \supset (Fx \equiv Bxt)]$$

which specifies that if x is sharply struck at any time t, then x is fragile if and only if x breaks at t.

Our earlier difficulty is now avoided, for if a is a nonfragile object that is never sharply struck, then that expression in R which follows the quantifiers is true of a; but this does not imply that 'Fa' is true. But the reduction sentence R specifies the meaning of 'F' only for application to those objects which meet the "test condition" of being sharply struck at some time; for these it states that fragility then amounts to breaking. For objects that fail to meet the test condition, the meaning of 'F' is left undetermined. In this sense, reduction sentences have the character of partial or conditional definitions.

Reduction sentences provide a satisfactory interpretation of the experiential import of a large class of disposition terms and permit a more adequate formulation of so-called operational definitions, which, in general, are not complete definitions at all. These considerations

---

[12] On this subject, see for example Langford (1941); Lewis (1946), pp. 210–30; Chisholm (1946); Goodman (1947); Reichenbach (1947), Chapter VIII; Hempel and Oppenheim (1948), Part III; Popper (1949); and especially Goodman's further analysis (1955). * See also the discussion by Scheffler, pp. 74–82. —Ed.

[13] Cf. Carnap, loc. cit. note 11. For a brief elementary presentation of the main idea, see Carnap (1938), Part III. The sentence R here formulated for the predicate 'F' illustrates only the simplest type of reduction sentence, the so-called bilateral reduction sentence.

suggest a greatly liberalized alternative to the requirement of definability:

(3.2) *The Requirement of Reducibility.* Every term with empirical significance must be capable of introduction, on the basis of observation terms, through chains of reduction sentences.

This requirement is characteristic of the liberalized versions of positivism and physicalism which, since about 1936, have superseded the older, overly narrow conception of a full definability of all terms of empirical science by means of observables,[14] and it avoids many of the shortcomings of the latter. Yet, reduction sentences do not seem to offer an adequate means for the introduction of the central terms of advanced scientific theories, often referred to as theoretical constructs. This is indicated by the following considerations: A chain of reduction sentences provides a necessary and a sufficient condition for the applicability of the term it introduces. (When the two conditions coincide, the chain is tantamount to an explicit definition.) But now take, for example, the concept of length as used in classical physical theory. Here, the length in centimeters of the distance between two points may assume any positive real number as its value; yet it is clearly impossible to formulate, by means of observation terms, a sufficient condition for the applicability of such expressions as 'having a length of $\sqrt{2}$ cm' and 'having a length of $\sqrt{2} + 10^{-100}$ cm'; for such conditions would provide a possibility for discrimination, in observational terms, between two lengths which differ by only 10^{-100} cm.[15]

It would be ill-advised to argue that for this reason, we ought to permit only such values of the magnitude, length, as permit the statement of sufficient conditions in terms of observables. For this would rule out, among others, all irrational numbers and would prevent us from assigning, to the diagonal of a square with sides of length 1, the length $\sqrt{2}$, which is required by Euclidean geometry. Hence, the principles of Euclidean geometry would not be universally applicable in physics. Similarly, the principles of the calculus would become inapplicable, and the system of scientific theory as we know it today would be reduced to a clumsy, unmanageable torso. This, then, is no way of meeting the difficulty. Rather, we shall have to analyze more closely the function of constructs in scientific theories, with a view to obtaining through such an analysis a more adequate characterization of cognitively significant terms.

Theoretical constructs occur in the formulation of scientific theories. These may be conceived of, in their advanced stages, as being stated in

---

[14] Cf. the analysis in Carnap (1936–37), especially section 15; also see the briefer presentation of the liberalized point of view in Carnap (1938).

[15] (Added in 1964.) This is not strictly correct. For a more circumspect statement, see note 12 in "A Logical Appraisal of Operationism" and the fuller discussion in section 7 of the essay "The Theoretician's Dilemma."

the form of deductively developed axiomatized systems. Classical mechanics, or Euclidean or some Non-Euclidean form of geometry in physical interpretation, present examples of such systems. The extralogical terms used in a theory of this kind may be divided, in familiar manner, into primitive or basic terms, which are not defined within the theory, and defined terms, which are explicitly defined by means of the primitives. Thus, for example, in Hilbert's axiomatization of Euclidean geometry, the terms 'point', 'straight line', 'between' are among the primitives, while 'line segment', 'angle', 'triangle', 'length' are among the defined terms.* The basic and the defined terms together with the terms of logic constitute the vocabulary out of which all the sentences of the theory are constructed. The latter are divided, in an axiomatic presentation, into primitive statements (also called postulates or basic statements) which, in the theory, are not derived from any other statements, and derived ones, which are obtained by logical deduction from the primitive statements.

From its primitive terms and sentences, an axiomatized theory can be developed by means of purely formal principles of definition and deduction, without any consideration of the empirical significance of its extralogical terms. Indeed, this is the standard procedure employed in the axiomatic development of uninterpreted mathematical theories such as those of abstract groups or rings or lattices, or any form of pure (i.e., noninterpreted) geometry.

However, a deductively developed system of this sort can constitute a scientific theory only if it has received an empirical interpretation[16] which renders it relevant to the phenomena of our experience. Such interpretation is given by assigning a meaning, in terms of observables, to certain terms or sentences of the formalized theory. Frequently, an interpretation is given not for the primitive terms or statements but rather for some of the terms definable by means of the primitives, or for some of the sentences deducible from the postulates.[17] Further-

---

* David Hilbert, *Foundations of Geometry*, transl. E. J. Townsend (La Salle, Ill.: The Open Court Publishing Co., 1962). —Ed.

[16] The interpretation of formal theories has been studied extensively by Reichenbach, especially in his pioneer analyses of space and time in classical and in relativistic physics. He describes such interpretation as the establishment of *coordinating definitions* (Zuordnungsdefinitionen) for certain terms of the formal theory. See, for example, Reichenbach (1928). More recently, Northrop [cf. (1947), Chap. VII, and also the detailed study of the use of deductively formulated theories in science, ibid., Chaps. IV, V, VI] and H. Margenau [cf., for example, (1935)] have discussed certain aspects of this process under the title of *epistemic correlation*.

[17] A somewhat fuller account of this type of interpretation may be found in Carnap (1939), §24. The articles by Spence (1944) and by MacCorquodale and Meehl (1948) provide enlightening illustrations of the use of theoretical constructs in a field outside that of the physical sciences, and of the difficulties encountered in an attempt to analyze in detail their function and interpretation.

more, interpretation may amount to only a partial assignment of meaning. Thus, for example, the rules for the measurement of length by means of a standard rod may be considered as providing a *partial* empirical interpretation for the term 'the length, in centimeters, of interval *i*', or alternatively, for some sentences of the form 'the length of interval *i* is *r* centimeters'. For the method is applicable only to intervals of a certain medium size, and even for the latter it does not constitute a full interpretation since the use of a standard rod does not constitute the only way of determining length: various alternative procedures are available involving the measurement of other magnitudes which are connected, by general laws, with the length that is to be determined.

This last observation, concerning the possibility of an indirect measurement of length by virtue of certain laws, suggests an important reminder. It is not correct to speak, as is often done, of "the experiential meaning" of a term or a sentence in isolation. In the language of science, and for similar reasons even in pre-scientific discourse, a single statement usually has no experiential implications. A single sentence in a scientific theory does not, as a rule, entail any observation sentences; consequences asserting the occurrence of certain observable phenomena can be derived from it only by conjoining it with a set of other, subsidiary, hypotheses. Of the latter, some will usually be observation sentences, others will be previously accepted theoretical statements. Thus, for example, the relativistic theory of the deflection of light rays in the gravitational field of the sun entails assertions about observable phenomena only if it is conjoined with a considerable body of astronomical and optical theory as well as a large number of specific statements about the instruments used in those observations of solar eclipses which serve to test the hypothesis in question.

Hence, the phrase, 'the experiential meaning of expression *E*' is elliptical: What a given expression "means" in regard to potential empirical data is relative to two factors, namely:

    I. *the linguistic framework L* to which the expression belongs. Its rules determine, in particular, what sentences—observational or otherwise—may be inferred from a given statement or class of statements;

    II. the theoretical context in which the expression occurs, that is, the class of those statements in *L* which are available as subsidiary hypotheses.

Thus, the sentence formulating Newton's law of gravitation has no experiential meaning by itself; but when used in a language whose logical apparatus permits the development of the calculus, and when combined with a suitable system of other hypotheses—including sentences which connect some of the theoretical terms with observation terms and thus establish a partial interpretation—then it has a bearing on observable phenomena in a large variety of fields. Analogous considerations are applicable to the term 'gravitational field', for example. It can be considered as having experiential meaning only within the context of a theory, which must be at least partially interpreted; and the experiential meaning of the term—as expressed, say, in the form of operational criteria for its application—will depend again on the theoretical system at hand, and on the logical characteristics of the language within which it is formulated.

4. COGNITIVE SIGNIFICANCE AS A CHARACTERISTIC OF INTERPRETED SYSTEMS

The preceding considerations point to the conclusion that a satisfactory criterion of cognitive significance cannot be reached through the second avenue of approach here considered, namely by means of specific requirements for the terms which make up significant sentences. This result accords with a general characteristic of scientific (and, in principle, even pre-scientific) theorizing: Theory formation and concept formation go hand in hand; neither can be carried on successfully in isolation from the other.

If, therefore, cognitive significance can be attributed to anything, then only to entire theoretical systems formulated in a language with a well-determined structure. And the decisive mark of cognitive significance in such a system appears to be the existence of an interpretation for it in terms of observables. Such an interpretation might be formulated, for example, by means of conditional or biconditional sentences connecting nonobservational terms of the system with observation terms in the given language; the latter as well as the connecting sentences may or may not belong to the theoretical system.

But the requirement of partial interpretation is extremely liberal; it is satisfied, for example, by the system consisting of contemporary physical theory combined with some set of principles of speculative metaphysics, even if the latter have no empirical interpretation at all. Within the total system, these metaphysical principles play the role of what K. Reach and also O. Neurath liked to call *isolated sentences:* They are neither purely formal truths or falsehoods, demonstrable or refutable by means of the logical rules of the given language system; nor do they have any experiential bearing; that is, their omission from the theoretical system would have no effect on its explanatory and predictive power in regard to potentially observable phenomena (i.e., the kind of phenomena described by observation sentences). Should we not, therefore, require that a cognitively significant system contain no isolated sentences? The following criterion suggests itself:

(4.1) A theoretical system is cognitively significant if and only if it is partially interpreted to at least such an extent that none of its primitive sentences is isolated.

But this requirement may bar from a theoretical system certain sentences which might well be viewed as permissible and indeed desirable. By way of a simple illustration, let us assume that our theoretical system T contains the primitive sentence

($S1$) $\qquad\qquad (x)[P_1x \supset (Qx \equiv P_2x)]$

where 'P_1' and 'P_2' are observation predicates in the given language L, while 'Q' functions in T somewhat in the manner of a theoretical construct and occurs in only one primitive sentence of T, namely $S1$. Now $S1$ is not a truth or falsehood of formal logic; and furthermore, if $S1$ is omitted from the set of primitive sentences of T, then the resulting system, T', possesses exactly the same systematic, that is, explanatory and predictive, power as T. Our contemplated criterion would therefore qualify $S1$ as an isolated sentence which has to be eliminated—excised by means of Occam's razor, as it were—if the theoretical system at hand is to be cognitively significant.*

But it is possible to make a much more liberal view of $S1$ by treating it as a partial definition for the theoretical term 'Q'. Thus conceived, $S1$ specifies that in all cases where the observable characteristic P_1 is present, 'Q' is applicable if and only if the observable characteristic P_2 is present as well. In fact, $S1$ is an instance of those partial, or conditional, definitions which Carnap calls bilateral reduction sentences. These sentences are explicitly qualified by Carnap as analytic (though not, of course, as truths of formal logic), essentially on the ground that all their consequences which are expressible by means of observation predicates (and logical terms) alone are truths of formal logic.[18]

Let us pursue this line of thought a little further. This will lead us to some observations on analytic sentences and then back to the question of the adequacy of (4.1).

Suppose that we add to our system T the further sentence

($S2$) $\qquad\qquad (x)[P_3x \supset (Qx \equiv P_4x)]$

where 'P_3', 'P_4' are additional observation predicates. Then, on the view that "every bilateral reduction sentence is analytic",[19] $S2$ would be analytic as well as $S1$. Yet, the two sentences jointly entail non-analytic consequences which are expressible in terms of observation predicates alone, such as[20]

---

* Occam's razor is the principle of eliminating unnecessary entities. —Ed.
[18] Cf. Carnap (1936–37), especially sections 8 and 10.
[19] Carnap (1936–37), p. 452.
[20] The sentence O is what Carnap calls the *representative sentence* of the couple consisting of the sentences $S1$ and $S2$; see (1936–37), pp. 450–53.

(O) $(x)[\sim (P_1x \cdot P_2x \cdot P_3x \cdot \sim P_4x) \cdot \sim (P_1x \cdot \sim P_2x \cdot P_3x \cdot P_4x)]$

But one would hardly want to admit the consequence that the conjunction of two analytic sentences may be synthetic. Hence if the concept of analyticity can be applied at all to the sentences of interpreted deductive systems, then it will have to be relativized with respect to the theoretical context at hand. Thus, for example, $S1$ might be qualified as analytic relative to the system T, whose remaining postulates do not contain the term 'Q', but as synthetic relative to the system T enriched by $S2$. Strictly speaking, the concept of analyticity has to be relativized also in regard to the rules of the language at hand, for the latter determine what observational or other consequences are entailed by a given sentence. This need for at least a twofold relativization of the concept of analyticity was almost to be expected in view of those considerations which required the same twofold relativization for the concept of experiential meaning of a sentence.

If, on the other hand, we decide not to permit $S1$ in the role of a partial definition and instead reject it as an isolated sentence, then we are led to an analogous conclusion: Whether a sentence is isolated or not will depend on the linguistic frame and on the theoretical context at hand: While $S1$ is isolated relative to T (and the language in which both are formulated), it acquires definite experiential implications when T is enlarged by $S2$.

Thus we find, on the level of interpreted theoretical systems, a peculiar rapprochement, and partial fusion, of some of the problems pertaining to the concepts of cognitive significance and of analyticity: Both concepts need to be relativized; and a large class of sentences may be viewed, apparently with equal right, as analytic in a given context, or as isolated, or nonsignificant, in respect to it.

In addition to barring, as isolated in a given context, certain sentences which could just as well be construed as partial definitions, the criterion (4.1) has another serious defect. Of two logically equivalent formulations of a theoretical system it may qualify one as significant while barring the other as containing an isolated sentence among its primitives. For assume that a certain theoretical system $T1$ contains among its primitive sentences S', S'', . . . exactly one, S', which is isolated. Then $T1$ is not significant under (4.1). But now consider the theoretical system $T2$ obtained from $T1$ by replacing the two first primitive sentences, S', S'', by one, namely their conjunction. Then, under our assumptions, none of the primitive sentences of $T2$ is isolated, and $T2$, though equivalent to $T1$, is qualified as significant by (4.1). In order to do justice to the intent of (4.1), we would therefore have to lay down the following stricter requirement:

(4.2) A theoretical system is cognitively significant if and only if it is partially interpreted to such an extent that in no system equivalent to it at least one primitive sentence is isolated.

Let us apply this requirement to some theoretical system whose postulates include the two sentences $S1$ and $S2$ considered before, and whose other postulates do not contain 'Q' at all. Since the sentences $S1$ and $S2$ together entail the sentence O, the set consisting of $S1$ and $S2$ is logically equivalent to the set consisting of $S1$, $S2$ and O. Hence, if we replace the former set by the latter, we obtain a theoretical system equivalent to the given one. In this new system, both $S1$ and $S2$ are isolated since, as can be shown, their removal does not affect the explanatory and predictive power of the system in reference to observable phenomena. To put it intuitively, the systematic power of $S1$ and $S2$ is the same as that of O. Hence, the original system is disqualified by (4.2). From the viewpoint of a strictly sensationalist positivism as perhaps envisaged by Mach, this result might be hailed as a sound repudiation of theories making reference to fictitious entities, and as a strict insistence on theories couched exclusively in terms of observables. But from a contemporary vantage point, we shall have to say that such a procedure overlooks or misjudges the important function of constructs in scientific theory: The history of scientific endeavor shows that if we wish to arrive at precise, comprehensive, and well-confirmed general laws, we have to rise above the level of direct observation. The phenomena directly accessible to our experience are not connected by general laws of great scope and rigor. Theoretical constructs are needed for the formulation of such higher-level laws. One of the most important functions of a well-chosen construct is its potential ability to serve as a constituent in ever new general connections that may be discovered; and to such connections we would blind ourselves if we insisted on banning from scientific theories all those terms and sentences which could be "dispensed with" in the sense indicated in (4.2). In following such a narrowly phenomenalistic or positivistic course, we would deprive ourselves of the tremendous fertility of theoretical constructs, and we would often render the formal structure of the expurgated theory clumsy and inefficient.

Criterion (4.2), then, must be abandoned, and considerations such as those outlined in this paper seem to lend strong support to the conjecture that no adequate alternative to it can be found; that is, that it is not possible to formulate general and precise criteria which would separate those partially interpreted systems whose isolated sentences might be said to have a significant function from those in which the isolated sentences are, so to speak, mere useless appendages.

We concluded earlier that cognitive significance in the sense intended by recent empiricism and operationism can at best be attributed to sentences forming a theoretical system, and perhaps rather to such systems as wholes. Now, rather than try to replace (4.2) by some alternative, we will have to recognize further that cognitive significance in a system is a matter of degree: Significant systems range from those whose entire extralogical vocabulary consists of observation terms, through theories whose formulation relies heavily on theoretical constructs, on to systems with hardly any bearing on potential empirical findings. Instead of dichotomizing this array into significant and nonsignificant systems it would seem less arbitrary and more promising to appraise or compare different theoretical systems in regard to such characteristics as these:

   a. the clarity and precision with which the theories are formulated, and with which the logical relationships of their elements to each other and to expressions couched in observational terms have been made explicit;
   b. the systematic, that is, explanatory and predictive, power of the systems in regard to observable phenomena;
   c. the formal simplicity of the theoretical system with which a certain systematic power is attained;
   d. the extent to which the theories have been confirmed by experiential evidence.

Many of the speculative philosophical approaches to cosmology, biology, or history, for example, would make a poor showing on practically all of these counts and would thus prove no matches to available rival theories, or would be recognized as so unpromising as not to warrant further study or development.

If the procedure here suggested is to be carried out in detail, so as to become applicable also in less obvious cases, then it will be necessary, of course, to develop general standards, and theories pertaining to them, for the appraisal and comparison of theoretical systems in the various respects just mentioned. To what extent this can be done with rigor and precision cannot well be judged in advance. In recent years, a considerable amount of work has been done towards a definition and theory of the concept of degree of confirmation, or logical probability, of a theoretical system;[21] and several contributions have been made towards the clarification of some of the other ideas referred to above.[22] The continuation of this research represents a challenge for further constructive work in the logical and methodological analysis of scientific knowledge.

[21] Cf., for example, Carnap (1945)1 and (1945)2, and especially (1950). Also see Helmer and Oppenheim (1945).
[22] On simplicity, cf. especially Popper (1935), Chap. V; Reichenbach (1938), §42; Goodman (1949)1, (1949)2, (1950); on explanatory and predictive power, cf. Hempel and Oppenheim (1948), Part IV.

REFERENCES

AYER, A. J., *Language, Truth and Logic*, London, 1936; 2nd ed. 1946.

CARNAP, R., "Testability and Meaning," *Philosophy of Science*, 3 (1936) and 4 (1937).

CARNAP, R., "Logical Foundations of the Unity of Science," in: *International Encyclopedia of Unified Science*, I, 1; Chicago, 1938.

CARNAP, R., *Foundations of Logic and Mathematics*, Chicago, 1939.

CARNAP, R., "On Inductive Logic," *Philosophy of Science*, 12 (1945). Referred to as (1945)1 in this article.

CARNAP, R., "The Two Concepts of Probability," *Philosophy and Phenomenological Research*, 5 (1945). Referred to as (1945)2 in this article.

CARNAP, R., *Logical Foundations of Probability*, Chicago, 1950.

CHISHOLM, R. M., "The Contrary-to-Fact Conditional," *Mind*, 55 (1946).

CHURCH, A., Review of Ayer (1946), *The Journal of Symbolic Logic*, 14 (1949), 52–53.

FEIGL, H., "Existential Hypotheses: Realistic vs. Phenomenalistic Interpretations," *Philosophy of Science*, 17 (1950).

GOODMAN, N., "The Problem of Counterfactual Conditionals," *The Journal of Philosophy*, 44 (1947).

GOODMAN, N., "The Logical Simplicity of Predicates," *The Journal of Symbolic Logic*, 14 (1949). Referred to as (1949)1 in this article.

GOODMAN, N., "Some Reflections on the Theory of Systems," *Philosophy and Phenomenological Research*, 9 (1949). Referred to as (1949)2 in this article.

GOODMAN, N., "An Improvement in the Theory of Simplicity," *The Journal of Symbolic Logic*, 15 (1950).

GOODMAN, N., *Fact, Fiction, and Forecast*, Cambridge, Massachusetts, 1955.

HELMER, O. and P. OPPENHEIM, "A Syntactical Definition of Probability and of Degree of Confirmation." *The Journal of Symbolic Logic*, 10 (1945).

HEMPEL, C. G. and P. OPPENHEIM, "Studies in the Logic of Explanation," *Philosophy of Science*, 15 (1948). (Reprinted in this volume.)

LANGFORD, C. H., Review in *The Journal of Symbolic Logic*, 6 (1941), 67–68.

LEWIS, C. I., *An Analysis of Knowledge and Valuation*, La Salle, Ill., 1946.

MACCORQUODALE, K. and P. E. MEEHL, "On a Distinction Between Hypothetical Constructs and Intervening Variables," *Psychological Review*, 55 (1948).

MARGENAU, H., "Methodology of Modern Physics," *Philosophy of Science*, 2 (1935).

NORTHROP, F. S. C., *The Logic of the Sciences and the Humanities*, New York, 1947.

O'CONNOR, D. J., "Some Consequences of Professor A. J. Ayer's Verification Principle," *Analysis*, 10 (1950).

PAP, A., *Elements of Analytic Philosophy*, New York, 1949.

POPPER, K., *Logik der Forschung*, Wien, 1935.

POPPER, K., "A Note on Natural Laws and So-Called 'Contrary-to-Fact Conditionals'," *Mind*, 58 (1949).

REICHENBACH, H., *Philosophie der Raum-Zeit-Lehre*, Berlin, 1928.

REICHENBACH, H., *Elements of Symbolic Logic*, New York, 1947.

RUSSELL, B., *Human Knowledge*, New York, 1948.

SCHLICK, M., "Meaning and Verification," *Philosophical Review*, 45 (1936). Also reprinted in Feigl, H. and W. Sellars, (eds.) *Readings in Philosophical Analysis*, New York, 1949.

SPENCE, KENNETH W., "The Nature of Theory Construction in Contemporary Psychology," *Psychological Review*, 51 (1944).

ISRAEL SCHEFFLER

Prospects of a
Modest Empiricism, I

INTRODUCTION [1]

The heart of modern empiricism has been its doctrine of empirical meaning, with its sharp line between the verifiable and the unverifiable and its rejection of non-analytic,[2] non-experiential statements as nonsense. This doctrine has, however, fallen on evil days. Increasing logical precision in philosophy has weakened rather than strengthened it, while advanced theoretical physics seems more and more a living counter-example, to be accommodated only by stretching the doctrine out of all shape. Once a proud polemical tool, the doctrine has thus come to be treated as a problem or a proposal. Yet the label "empiricist" continues to function as a philosophical symbol, and it is not sufficiently realized that the decline of the doctrine calls for new examination of the label. Indeed, it is not too much to claim that without such new examination, the idea of empiricism may be thought empty; or, shall we say, meaningless.

Reprinted and excerpted from the *Review of Metaphysics*, Vol. 10 (1956), 383, 602–625, by kind permission of the author and publisher.

[1] I wish to thank Professors C. G. Hempel, N. Goodman, and N. Chomsky for their criticisms of an earlier draft of this paper and for helpful discussions of related problems.

[2] "Analytic," as used throughout the present paper, embraces both analytically true and analytically false statements.

IV PROBLEMS OF A MODEST EMPIRICISM: DISPOSITIONAL AND TRANSCENDENTAL TERMS

1-2. The Problem of Disposition Terms[3]

We may begin by characterizing the basic logical apparatus of E as comprising joint or alternative denial,* universal quantification, and overlapping[4] or membership, with suitable rules for sentence formation. Next, we may provide a list of (observational) predicates which are clearly acceptable, that is, for which we recognize clear applications. Our rules will guarantee the inclusion of all modes of quantification, will guarantee the sentencehood of all denials of sentences in E, while our vocabulary and rules will exclude sentences clearly recognizable as meaningless presystematically. Most recent discussions have followed Carnap[5] in treating so-called dispositional predicates as a special problem. Hempel's treatment,[6] for instance, holds that the language just described, with all extralogical predicates clearly observational, is inadequate, since it leaves no room for dispositional terms not definable within it. Thus, for example, "is magnetic," "is at temperature 100°C," "is irritable," and so forth, though predicable of observable entities, and important for everyday and scientific discourse, are not definable by those observation-predicates describing usual test-operations determining their applicability. Why this is so may be seen from the following examples, incorporating Carnap's well-known arguments. Suppose we wish to define "x weighs 5 lbs. at time t" by "If x is placed on scale S at t, S's pointer moves to '5' at t." This proposed definiens, formulated as a material conditional, is equivalent to "x is not placed on scale S at t, or S's pointer moves to '5' at t." But clearly the latter (and hence, by equivalence, the former) open sentence is true of every object o, taken as value of "x," such that o is not placed on scale S at t. Hence this definition would assign a weight of 5 lbs. at t to, among other things, Mt. Vesuvius, the Eiffel Tower, and every ant not on S at t.

To replace the material conditional of the definiens by a subjunctive

[8] For the first part of this study see this *Review*, X (March 1957), 383–400.

* The joint denial of A and B is the sentence which is true when neither A nor B is true; the alternative denial of A and B is the sentence which is true when either A or B is true but not both. If one takes either of these sentence forming operators as primitive the other operators (and, or, equivalence, etc.) can be defined.

[4] N. Goodman, *The Structure of Appearance* (Cambridge, Mass., 1961), pp. 42–55.

[5] R. Carnap, "Testability and Meaning," *Philosophy of Science*, III (1936) and IV (1937).

[6] C. G. Hempel, *Fundamentals of Concept Formation in Empirical Science* (Chicago, 1952), pp. 23–29.

locution; that is, "If x were placed on scale S at t, S's pointer would move to '5' at t" is indeed intuitively more adequate, but such a locution is itself not included in the repertoire of E as thus far described, while its reducibility to this repertoire has not been shown. Thus if dispositional terms cannot be directly defined in E, their subjunctive interpretation is simply irrelevant, at present, to their reduction to E.

Carnap's notion of reduction-sentences has been widely acclaimed as a solution to the problem of introducing disposition terms into E. Illustrating this notion by reference to our present example, we would replace our whole definitional equivalence by the (bilateral) reduction-sentence, "If x is placed on scale S at t, then x weights 5 lbs. at t if and only if S's pointer moves to '5' at t." The difficulty originally encountered by our first proposed definition is here successfully avoided. For, though the whole reduction-sentence is indeed true of every object o not placed on S at t, for example, Mt. Vesuvius, we cannot therefrom infer that every such o weighs 5 lbs. at t.

Certain difficulties, however, soon become apparent. Let "T" stand in place of some predicate of E applying to x when and only when x fulfills certain test conditions, let "R" stand in place of another predicate of E applicable to x when and only when x exhibits a specified reaction, and let "D" stand in place of the dispositional term to be introduced. It is clear that in:

(R1) $$Tx \supset (Dx \equiv Rx)$$

"Tx" cannot be universally false, and that in:

(R2) $$Tx \supset (Rx \supset Dx)$$

"$Tx \cdot Rx$" must be true of some x, for [assuming that in each case (R1) or (R2) is offered as the sole reduction-sentence for "D"] if this were not so, the application of "D" would remain undetermined for every x. But nevertheless, (R1) or (R2) can be fulfilled trivially by some would-be predicate otherwise objectionable.

For example, "is spiritually ectoplasmic" will presumably not be reducible into E because applying to no x, but "is a paper clip and is in desk d at t or is not a paper clip and is spiritually ectoplasmic" is reducible by (R1) or (R2) if we put "is a paper clip" in place of "T" and "is in desk d at t" in place of "R." From the point of view of someone who wants to use "is spiritually ectoplasmic" freely in purported description of certain human beings, the reducible would-be predicate will serve equally well and hardly represents a concession. Moreover, though eliminable in one trivial context, it is ineliminable in all other contexts, notably those in which such a person was anxious to use "is spiritually ectoplasmic" in the first place, and for which we deemed such use objectionable, that is, in application to some non-paper-clips.

Of course, introduction via (R1) or (R2) does not positively specify a use for our objectionable term in these wider contexts either. But then, if we care only about the narrower use which it does determine, we are also able to *define* such use, in effect replacing the objectionable "D" by "is a paper clip and is in desk d at t," a point recently made by Goodman." [7] To replace such definition by reduction-sentences in the interests of the wider use of dispositional terms like "is magnetic" opens the door also to objectionable terms of the kind above exemplified, and amounts then to their bald acceptance as primitives. If, moreover, our claim is that for specific terms like "is magnetic" we do, as a matter of common or scientific procedure, have a use wider than that specifiable by such other predicates as are included in E, then we may add such terms to our list of E's primitives. Such piecemeal additions would not simultaneously admit objectionable terms, as would wholesale sanction of the method of reduction sentences.

In support of the above treatment of disposition terms, the following considerations may be adduced:

(a) We have noted the looseness of the general term "is an observational predicate" and the need for supplanting its use by a finite listing in any rigorous treatment. Included in such a listing would be sufficiently clear terms with relatively determinate applications to observable entities. Now it is worth noting that the disposition terms under discussion are all predicates of our initially chosen observable entities, that is, applicable to nothing outside the range of the variables already specified for E as suitable for its initial, observational predicates. Thus, if the latter are characterized by relatively determinate application to entities within this range, the same may hold for so-called dispositional predicates, especially in view of the fact that determinateness of application is a continuous matter, a question of degree. Furthermore, unless such determinateness did characterize some dispositional terms in contexts wider than those represented by customarily stated test-conditions, we could hardly charge E with inadequacy for omitting them.[8] There is, thus, no theoretical ground for

---

[7] N. Goodman, *Fact, Fiction, and Forecast* (Cambridge, Mass., 1955), p. 50.

[8] That is, inadequacy in descriptiveness, or in capacity for formulating evidence-statements. Certain so-called dispositional terms might be defended on much the same grounds as theoretical terms, however, in a manner to be discussed in later sections, and involving no requirement for reduction-sentences. Such defense must, of course, be treated differently, but my aim here is to counter the prevalent idea that partial specification of the meaning of disposition terms by reduction sentences creates a special intermediate class of terms with unique functions in formulating

denying to *every* dispositional term the status of observational predicate.

(b) It is also worth remembering that dispositionality is not absolute. Most physical examples recently discussed, for example, "is soluble at *t*," "is magnetic at *t*," are contrasted with ostensible non-dispositional terms, for example, "dissolves at *t*," "is placed near iron fillings at *t*," which are themselves physical-object terms. But the latter are often treated as in effect dispositional by phenomenalistically-minded thinkers: consider, for example, Mills' doctrine[9] of matter as permanent possibility of sensation, or Lewis' terminating judgments as expressing the content of objective beliefs.[10] The converse is, also, not unthinkable. If no term is dispositional in an absolute sense, we need not balk at taking some terms customarily labelled "dispositional" as observational predicates in *E*.

(c) Much of the puzzlement over dispositional terms seems to be a product of platonist semantics, at least in part. Predicates are taken to designate properties, in addition to denoting each entity to which they apply. Properties designated by non-dispositional terms are then said to be themselves observable, while those designated by dispositional terms are said to be not observable, or, at least, not observable to the same degree or in the same sense. To drop platonist semantics is thus to dispose of part of the puzzlement over dispositional terms. For, left only with denotation, we must admit that dispositional as well as non-dispositional terms apply to observable entities equally, and without sharp distinction as to determinateness or vagueness. No further question about the relative observability of *properties* remains.

Even had we kept our platonist semantics, it would have been hard to explicate the notion of observability of properties in such a way as to render dispositional properties clearly unobservable in some relevant sense. For, to take just one point, complex properties, each component of which is observable, are presumably to be considered themselves

---

our information about observables, i.e., capable of *continued* adequacy in expressing such information with the *continuing growth* of scientific investigation. Either a term has a clear enough denotative use to be observationally defined or to serve as an observational primitive (in which case the method of reduction-sentences is unnecessary) or else it must be justified on the same grounds as theoretical, non-observational terms (in which case the method of reduction-sentences is also unnecessary). See Goodman, *Fact, Fiction and Forecast*, p. 60, n. 11, for criticism of reduction as a kind of definition.

[9] J. S. Mill, *An Examination of Sir William Hamilton's Philosophy*, 6th ed. (London, 1889) p. 225 ff., esp. p. 233.

[10] *An Analysis of Knowledge and Valuation* (LaSalle, Ill., 1946).

observable. But then we cannot by simple inspection rule out the contingency that a given dispositional property may turn out equivalent to some such complex property, and hence also observable. This and other difficulties arise in analogous forms if, having dropped platonism, we attempt to define some notion of observability or descriptiveness which shall distinguish between certain predicates (as applicable to things in virtue of certain specific sensory qualities) and others, though all apply to observable entities within the same range. This attempt is often made not for the purpose of distinguishing dispositional from non-dispositional predicates, but rather in order to distinguish ethical from non-ethical terms, in which case it seems to me clearly unworkable.[11]

(d) It has been noted under (b) above, that dispositionality is not absolute, but depends on the system chosen. It is now important to see that even within a given system, a predicate P may be dispositional with respect to one predicate or set of predicates Q and not dispositional with respect to some other W, on the basis of which it is fully definable. There is then no point in taking P as ostensibly saying something about possibilities or potentialities quite generally. Thus, we may denote as "played" every record which, at any time, is actually put on the turntable of some record player of familiar type and, with the needle in position, produces recognized music or speech. Aside from these, we may denote as "playable" also certain other objects, for example, records accidentally shattered before ever reaching the turntable. Asked to explain the latter term, we should naturally try to do so by reference to the former via recourse to notions such as capability, possibility or potentiality, to the use of the subjunctive, or to devices like reduction-sentences. And each of the latter courses seems to point to a queerness inherent in the predicate "is playable"; it talks about something else than what is actually the case in our world. Consider, however, that everything in the above sense playable is a record of music or speech, meeting rather definite specifications as to form, shape, and history, and conversely, that every such record is playable. This means that "is playable" is definable in terms of those predicates by means of which such specifications are stated, without recourse to possibility or the subjunctive. Assuming that such predicates are among the primitive observational terms of a given system S, we now have a situation in which one predicate ("is playable") is dispositional with respect to another ("is played") in S, but is fully definable and non-dispositional with respect to certain others, also in S. The upshot is that, even for a given system, where a predicate P seems to be dis-

[11] I. Scheffler, "Anti-Naturalist Restrictions in Ethics," *Journal of Philosophy*, L (1953), 457–66, esp. 457–60.

positional in relation to some other predicate Q within the system, we cannot generally attribute some special reference to possibility to the predicate P as such, even as it figures within the given system.

(e) The previous four subsections are all concerned with breaking down the customary sharp dichotomy between dispositional and non-dispositional terms and with pointing out that any predicate having a sufficiently determinate application to chosen observable entities may qualify as a clearly observational predicate, and may even turn out definable by other specified observational terms. In line with this general purpose, it was pointed out that dispositional as well as non-dispositional predicates may apply to observable entities within the same range, as customarily construed, that giving up platonist semantics with its attendant obscurities, we have no clearer way of distinguishing degrees of observationality or descriptiveness for terms with equal applicability to observable entities; and that the attribution of dispositionality or non-dispositionality to any predicate P varies with the chosen system and with the predicates taken as standards. The conclusion as regards introducing so-called dispositional terms into our observational system E seems to be that each term must be judged on its merits, that any such term might be sufficiently clear to qualify as an observational primitive or get itself defined properly, and that if such a term is really needed in E for adequacy it will indeed be of this type.

Is there then no specific problem associated with disposition terms? Such a judgment would be wrong. There is an important problem here, but it is independent of the one with which we have been occupied above, namely, to see whether it is possible to construct an observational E, free from meaningless sentences and adequate for expressing our beliefs in specified domains, for example, the sciences. This other problem is to define the *relationship between* dispositional terms and what are customarily taken as their respective non-dispositional counterparts in specified contexts. More exactly, it is to define the semantic relative term "is the dispositional counterpart of," or, put otherwise, to define a dispositional operator, say "-ible," attachable to predicates so as to form their dispositional predicate counterparts. Goodman has recently discussed this as "the problem of characterizing a relation such that if the initial manifest predicate 'Q' stands in this relation to another manifest predicate or conjunction of manifest predicates 'A', then 'A' may be equated with the dispositional counterpart—'Q-able' or 'Q_d'—of the predicate 'Q'."

That this general problem is independent of any given decision on a particular dispositional term is pointed out by Goodman as follows. "Observe first that solution of the general problem will not automatically provide us with a definition for each dispositional predicate; we shall need additional special knowledge in order to find the auxiliary predicate that satisfies the general formula—that is, that is related in the requisite way to the initial manifest predicate. But on the other hand, discovery of a suitable definition for a given dispositional predicate need not in all cases wait upon solution of the general problem. If luck or abundant special information turns up a manifest predicate 'P' that we are confident coincides in its application with 'flexible,' we can use 'P' as definiens for 'flexible' without inquiring further about the nature of its connection with 'flexes' " (*Fact, Fiction and Forecast,* pp. 48–49). It has here been further argued that the decision to regard a term as dispositional varies, and that even where it is positive, it is no bar to observationality in any important sense. In particular, where application of a given term is fairly determinate in contexts wider than those represented by customarily stated test-situations (and hence where definition of a narrower term fails to do our *de facto* usage justice) no general reason can be given against taking such a term as observational, and pending possible definition in the course of a general attempt at economy, including it as primitive. Since such terms with wide determinate application are the only dispositional terms whose omission might seriously mar *E*'s adequacy, our initial problem can be affirmatively answered for crucial *dispositional* terms: if we add every such term required for *E*'s adequacy to its primitives, we do not in general decrease the observational character of *E*. This does not mean we should not try to define these terms anyway, but we should try to define as many terms as possible *generally,* to reduce our stock of primitive notions. Nor does this mean that we dispense with the *general problem* of dispositions earlier mentioned, since it is independent. Nor, finally, does this mean that every term *customarily taken* as dispositional will have such a determinate application as to be judged sufficient; for many it will no doubt be advisable to define narrower notions, as Goodman suggests; but these terms are precisely those whose omission represents no threat to *E*'s adequacy, since they have no well-determined, prior uses in wider contexts. It is worth noting that Goodman's example of the possibility of defining "flexible" implies a prior, well-determined application which may control our definition, and hence the correlative possibility of taking this term as observational. Unless, moreover, we were aware of such previous application prior to our attempt at definition, we should have no way of guiding ourselves in our attempt, nor any reason for considering such attempt worthwhile.

13. The Problem of Theoretical or Transcendental Terms

If disposition terms required for descriptive adequacy may be accommodated in *E* without marring its observational character, as we have above argued, we still face a major obstacle to *E*'s adequacy, that is, the case of so-called abstract, theoretical, or transcendental terms. These terms, unlike dispositional predicates, do not generally purport to apply to entities within the range of application of our clearly observational terms. They are typically non-observational and beyond the reach even of reduction sentences. It is generally claimed that they are required not because of their usefulness in expressing available observational evidence, but rather because, by their introduction in the context of certain developed theories, comprehensive relationships become expressible in desirable ways on the observable level.

Such theories seem to commit us to a new range of entities as values of variables to which their transcendental terms may be attached in existentially quantified statements, for example, "There is an electron," "Something is a positron." For the entities here required are not such as are qualifiable by our hitherto accepted observational predicates. We have, for example, no clearly true sentence such as "x is an electron and x is red," or "x is an electron and x is non-red," for any value of "x." Historically, this is perhaps related to the distinction between primary and secondary qualities, a distinction which, in some form, it is dangerous to overlook in the interpretation of modern scientific theories. Thus, it is by now well known that confusion results both from the popularization of advanced theories through pictorial description in the common language of observation, and from the ascription of exclusive reality to the entities presupposed by such of these theories as we deem true. But to maintain the relevant distinction among ranges seems clearly to mean the abandonment of even our modest, revised empiricism. For the predicates appropriate to one of these ranges are non-observational, and to admit the necessity of theories couched in such terms in order for E to be adequate is to admit that no adequate, purely observational E exists which is even a sufficient condition of cognitive significance. (Thus those who insist on some such distinction by saying that "theoretical entities" have only such properties as are attributed to them by their respective theoretical contexts are, if they let the matter rest here, abandoning even our modest version of empiricism.) Furthermore, aside from empiricism, the perpetuation of a distinction among ranges seems to have generally puzzling aspects, which have troubled philosophers of science recurrently: (a) If adequacy requires that clearly non-observational terms be eligible for admission into the language of science, then, since these include terms ordinarily deemed meaningless, *what* are we believing in committing ourselves to science? (b) If science explains by providing *true* premises from which the relevant problematic data may be derived, how can theories that are *meaningless* in the ordinary sense be said to explain? In short, *even if we are not interested in relating cognitive significance to observationality,* but are concerned with constructing some weakest language adequate for formulating some specified segment of our scientific beliefs, we may be troubled to find ourselves explicitly allowing clearly meaningless units to be built into our language structure, in some ordinary sense of "meaningless."

14. Pragmatism

The qualification embodied in the final phrase of the last sentence is a clue to one widespread attempt to cope with the troubles discussed: the view which takes the *system* to be the unit of significance, and adopts a wider notion of meaningfulness in deference to scientific practice. This view I shall label 'pragmatism.' The alternative reaction, which I shall discuss in later sections, I call 'fictionalism.' [12] Both views may be considered independently of the issue of empiricism, it seems to me, in view of the final set of remarks in the previous section. For, however we delimit the terms initially admissible (whether by reference to observation or not), we face the problem of interpreting all the others which seem to be required in increasing numbers with the theoretical development of science. Nevertheless, a consideration of this problem in abstraction from empiricism will bear rather directly on it as embodied in our revised form: pragmatism will negate this empiricism, while fictionalism will render it at least possible.

Pragmatism, then (in my terminology), accepts as fact that no initial listing of admissible terms is sufficient for the formulation of our scientific beliefs, and that the admission of any term is conceivable (hence legitimate) on the grounds of its utility in prediction and theoretical simplification. It admits, furthermore, that some such terms are meaningless, *in one usual sense.* But here it takes the bull by the horns, claiming that this sense is irrelevant for analyzing our scientific beliefs and practices. For any sense of "meaningless" which renders what is predictively useful meaningless is inadequate for philosophy of science, however relevant it may be in other contexts. If science finds, for example, transcendental theories fruitful within whole systematic con-

---

[12] The labels I introduce here and in following sections are related to, but not intended as names of, specific philosophies associated with familiar historical movements or individual thinkers. They refer rather to characteristic trends, somewhat oversimplified and idealized, perhaps, in comparison to actually held philosophies. Nevertheless, they are influential and salient trends and will be recognizable, I hope, as elements of much of the recent literature in philosophy of science and theory of knowledge. As recent illustrations (in a loose sense) of pragmatism see R. Carnap, "Empiricism, Semantics, and Ontology," *Revue Internationale de Philosophie*, XI (1950), reprinted in L. Linsky, *Semantics and the Philosophy of Language* (Urbana, Ill., 1952); and W. V. Quine, "Two Dogmas of Empiricism," *Philosophical Review* (1951), included in W. V. Quine; *From a Logical Point of View* (Cambridge, Mass., 1953); of instrumentalistic fictionalism, see Toulmin, S. E., *The Philosophy of Science* (London, 1953). Regarding vacillation between the latter two trends, see Nagel's discussion of Dewey in *Sovereign Reason* (Glencoe, Ill., 1954), pp. 110–115.

texts, our notions of cognitive significance must reflect this fact. If science introduces terms not only by reference to prescientific usage or explicit test-methods, but within the network of whole theoretical frameworks justified by their predictive utility in subsequent inquiry, this is a *bona fide* fact about cognitive significance, not a problem. We must, accordingly, for the pragmatists, admit the *significance* of whole systems with unrestricted vocabularies, provided that they are, at some points, functionally tied to our initially specified language. Since, however, this proviso excludes no term at all, and virtually no system at all (every system meets this requirement by addition of one conjunct in the initial language, and every term is part of some system meeting this requirement), pragmatism supplements it by stressing some simplicity factor which presumably is to eliminate certain systems, but which is not to be so stringent as to eliminate every system which overflows the bounds of the initially specified language. In no account is the treatment of simplicity very precise, but in some accounts it is intended as a matter of degree so that cognitive significance is broadened correspondingly. It is, further, not very clear how considerations of simplicity are to be applied in determination of *confirmedness* or *truth* as distinct from *significance*. Nevertheless, the pragmatist reaction to the problem of interpreting transcendental terms and then theoretical contexts is clearly to accept their fruitfulness and to defend, in consequence, a broader notion of significance applicable to whole systems. This systematic emphasis is supported also by inference to well-known analyses of testing which show the theoretical revisability of every segment of a system when any segment is ostensibly under review. A corollary of this pragmatist treatment is its insistence that questions of ontology are scientific questions, since it takes the range of significant ontological assertion to be solely a function of scientific utility in practice and denies all independent language restrictions based on intuitive clarity or observationality. And rejecting such independent restrictions, it solves the two initial difficulties noted above (at the end of section 13) by (a) denying that we believe meaningless assertions in committing ourselves to science and by (b) affirming the possible truth and explanatory power of transcendental theories.

To what extent is the pragmatist position in favor of a broader notion of significance positively supported by the arguments it presents? Its strong point is obviously its congruence with the *de facto* scientific use of transcendental theories and with the interdependence of parts of a scientific system undergoing test. These facts are, however, not in themselves *conclusive* evidence for significance, inasmuch as many kinds of things are used in science with no implication of cognitive significance, that is, truth-or-falsity; and many things are interdependent under scientific test without our feeling that they are therefore included within the cognitive system of our assertions. Clearly "is useful," "is fruitful," "is subject to modification under test," and so forth, are applicable also to non-linguistic entities, for example, telescopes and electronic computers. On the other hand, even linguistic units judged useful and controllable via empirical test may conceivably be construed as non-significant machinery, and such construction is not definitely ruled out by pragmatist arguments. This, even if we accept pragmatism's positive grounds, we *need* not broaden our original notion of literal significance. And it further follows that our revised empiricism is not refuted by pragmatism.

15. Pragmatism and Fictionalism*

But if not refuted, our empiricism remains beset with the problem of interpreting transcendental terms and theories. If pragmatism's positive grounds do not, that is, *establish* the literal significance of transcendental theories, it is not thereby demonstrated that they are eliminable or otherwise interpretable as non-beliefs, that is, mere instruments. Any view which takes them to be either I call 'fictionalism.' Clearly if fictionalism can show how transcendental terms are eliminable from our corpus of scientific beliefs, it will have removed transcendental theories from the domain of beliefs which need to be encompassed in *E*, and it will have destroyed a major obstacle to our revised empiricism. Short of showing eliminability, if fictionalism can plausibly construe transcendental theories as mere machinery without literal meaning, it will avoid the need for expressing such theories in *E*, and again make way for our revised empiricism.

16. Instrumentalistic Fictionalism

Perhaps the easiest, and by far the most popular type of fictionalism is one which simply disavows the belief-character of transcendental theories without claiming their eliminability from scientific discourse. Indeed such a fictionalism often goes with a positive indifference to the question of their eliminability, or even champions their ineliminability; we might aptly label this type 'instrumentalism,' and note in passing that some writers have vacillated between pragmatism and instrumentalism (in our present terminology) or have confused the two. Instrumentalistic fictionalism, then, holds that some scientific theories are not significant, but that they are moreover not intended as formulations of belief or as truths, being employed simply as mechanical devices for coordinating or generating *bona fide* assertions. Hence, again, transcendental theories are said to pose no problem; since they

* Scheffler's use of 'fictionalism' and 'instrumentalism' differs somewhat from that of the Introduction. —Ed.

do not represent *beliefs,* we need not worry about including them within any deliberate statement of our beliefs in some restricted language. Our problem, it will be recalled, was that clearly meaningless terms seem required for adequate expression of our scientific beliefs. Whereas pragmatism's answer is to deny that any terms usefully employed in science are meaningless in the relevant sense, fictionalism's answer is to deny that our objectionable terms are required for expression of *beliefs,* though they may be otherwise required. And our instrumentalistic variant supports this denial not by showing how to eliminate such terms from scientific language, but rather by stipulating how 'belief' is to be understood. Correspondingly, certain further stipulations are generally accepted as corollaries, for example, that transcendental theories be said to *hold* or *fail* rather than to be *true* or *false,* that they are adopted or abandoned rather than believed or denied, and so forth. Thus instrumentalism takes care of the difficulties mentioned (a) by insisting that we do not strictly *believe* but *hold* or *employ* some statements in science and (b) by generalizing the concept of explanation to allow such held theories to serve as explanatory grounds.

If pragmatism's positive grounds seemed to us unconvincing, instrumentalism's positive grounds seem to consist just in the intuitive meaningless of transcendental theories. But the point at issue is whether science requires us to believe such theories, and this point is not met but begged by arguing that the answer is negative since the theories are intuitively meaningless. We can, however, be more generous to both pragmatism and instrumentalism by taking them not as arguments but as decisions or resolutions: pragmatism's denial of the meaningless of transcendental theories represents a decision to apply to them the ordinary language of truth and falsity, and this, coupled with denial of the need for further interpretation, involves a rejection of even a modified empiricism, as we have above formulated it. Instrumentalism's denial of the belief-character of transcendental theories represents a decision to talk about such theories in different and special ways without any further changes. Taken as basic decisions, there would seem to be no way of refuting either position, and to this extent at least, ontology is independent of science. There is no way to refute the instrumentalist's denial of the belief-character of various theories which he continues to employ. We may charge his implicit conception of the nature of belief with being tenuous and merely verbal, and we may declare his disavowals of belief to be rather hollow unless he gives up using the sentences which he claims intellectually to disavow. Yet, if he sticks to his guns, and continues to remind us that we all *use* all kinds of objects which we hold meaningless, and feel no guilt upon reflection in continuing to use them, then he is secure. Ontology, then, is relative to the person, and independent of the used language. Just as our common use of available technology does not commit us all equally to the same beliefs, so our common use of scientific language does not dictate that we should all draw the same line between literal sense and nonsense therein.

Coming back to our modified empiricism now, it appears that if pragmatism, in choosing to deny it, does not thereby refute it, instrumentalism renders it trivial. For if the range of our *beliefs* is freely specifiable by intellectual decision independently of the *content of our discourse,* we can always guarantee E's adequacy by simply deciding to exclude recalcitrant sentences from this range. Our judgments of recalcitrance will, of course, vary; but one consistent with our modified empiricism is trivially always possible.

If, however, we interpret our modified empiricist problem more stringently and more objectively, that is, as not allowing for such a trivial answer, we must require of the empiricist fictionalist not simply that he appropriately adjust his terminology of belief but that he provide a method for eliminating transcendental terms and theories from scientific *discourse,* or of treating them within his discourse otherwise than as significant.

17. Syntactic Fictionalism

One such course open to the fictionalist is to provide a syntax for transcendental theories. Goodman and Quine[13] have in part thus dealt with the problem of treating mathematics nominalistically. Unable, at the time of their study, to translate all of mathematics into a nominalistic language, they developed a nominalistic syntax language enabling them to talk *about* and deal with the untranslated residue, thus *independently* supporting the claim that this residue could be treated as mere machinery without literal significance. Though, in one sense, they did not eliminate this residue, they did go considerably beyond a mere statement that it might be considered as machinery only. For they provided an alternative language without the (to them at that time) objectionable features of the original, such that it was capable of doing much the same job. As they put their view, "our position is that the formulas of platonistic mathematics are, like the beads of an abacus, convenient computational aids which need involve no question of truth. What is meaningful and true in the case of platonistic mathematics as in the case of the abacus is not the apparatus itself, but only the description of it: the rules by which it is constructed and run. These rules we do understand, in the strict sense that we can express

[13] N. Goodman and W. V. Quine, "Steps Toward a Constructive Nominalism," *Journal of Symbolic Logic,* XII (1947), 105–122.

them in purely nominalistic language. The idea that classical mathematics can be regarded as mere apparatus is not a novel one among nominalistically minded thinkers; but it can be maintained only if one can produce, as we have attempted to above, a syntax which is itself free from platonistic commitments. At the same time, every advance we can make in finding direct translations for familiar strings of marks will increase the range of the meaningful language at our command (p. 122)."

Such a syntactical approach has relevance far beyond the question of platonistic mathematics. It is in general open to the fictionalist who wishes to disavow the belief-character of some segment of received scientific discourse in more than the trivial sense discussed above in connection with instrumentalism. In the special case of nominalism, it was by no means initially obvious that a syntax could be constructed without platonistic features. Having such a syntax for mathematics, it seems possible to extend it to specified parts of empirical science by addition of predicates applicable to the extra-logical notation contained therein. In particular, such syntax could be developed for transcendental theories which the fictionalist cannot eliminate through translation but which he finds it objectionable to take as significant. For the non-nominalist who objects to taking as significant some particular transcendental theory, the task is, of course, much easier, for he has available all the tools of platonistic syntax. In a less trivial sense than that of instrumentalism, then, our modified empiricism may be feasible through syntactic construction. Note again, incidentally, that ontology turns out independent of our received scientific discourse, through the possibility of variable syntactic reinterpretation, and that only such elements as are needed for the applicability of syntactic predicates may be sufficient in the extreme case.

18. Eliminative Fictionalism

We may, finally, require our modified empiricism to show how transcendental terms may be eliminated from scientific discourse in favor of some other object-language discourse which is equivalent in some appropriate sense. Here it is well to recall that transcendental theories are justified generally as making possible the statement of comprehensive relationships (in desirable ways) on the observable level. Thus, if a way could be shown of appropriately stating these observational relationships in some theory, *S,* which otherwise differed from its transcendental counterpart only by lacking sentences with any transcendental term, *S* would be, in a reasonable sense, equivalent to that counterpart.

One such method is that of Craig, who states as one of his results,

". . . if K is any recursive set of non-logical (individual, function, predicate) constants, containing at least one predicate constant, then there exists a system whose theorems are exactly those theorems of T in which no constants other than those of K occur. In particular, suppose that T expresses a portion of a natural science, that the constants of K refer to things or events regarded as 'observable,' and that the other constants do not refer to 'observables' and hence may be regarded as 'theoretical' or 'auxiliary.' Then there exists a system which does not employ 'theoretical' or 'auxiliary' constants and whose theorems are the theorems of T concerning 'observables.' " [14]

Professor Hempel, discussing Craig's method, states concisely what is involved and what sense of "equivalence" is here relevant: "Craig's result shows that no matter how we select from the total vocabulary $V_{T'}$ of an interpreted theory T' a subset V_B of experiential or observational terms, the balance of $V_{T'}$, constituting the 'theoretical terms,' can always be avoided in sense (c)." This sense, which Hempel distinguishes from definability and translatability, he calls "functional replaceability" and describes as follows, "The terms of T might be said to be avoidable if there exists another theory T_B, couched in terms of V_B, which is 'functionally equivalent' to T in the sense of establishing exactly the same deductive connections between V_B sentences as T."

Professor Hempel offers, however, two reasons against the scientific use of Craig's method,* "no matter how welcome the possibility of such replacement may be to the epistemologist." One reason is that the functionally equivalent replacing system constructed by Craig's method "always has an infinite set of postulates, irrespective of whether the postulate set of the original theory is finite or infinite, and that his result cannot be essentially improved in this respect. . . . This means that the scientist would be able to avoid theoretical terms only at the price of forsaking the comparative simplicity of a theoretical system with a finite postulational basis, and of giving up a system of theoretical concepts and hypotheses which are heuristically fruitful and suggestive—in return for a practically unmanageable system based

---

[14] W. Craig, "On Axiomatizability Within a System," *Journal of Symbolic Logic,* XVIII (1953), 31, text and n. 9. See also W. Craig, "Replacement of Auxiliary Expressions," *Philosophical Review,* LXV (1956), 38–55.

* William Craig proved the following result: If we are given an axiomatized theory T whose nonlogical vocabulary is divided into two portions V_0 and V_1 then we can find a theory T' with infinitely many axioms such that the theorems of T' are exactly those theorems of T whose vocabulary consists entirely of V_0. A relatively nontechnical presentation of the proof and a discussion of its importance for philosophy of science may be found in "Craig's Theorem" by Hilary Putnam in the *Journal of Philosophy,* Vol. 62 (1965), 251–60.

upon an infinite, though effectively specified, set of postulates in observational terms." [15]

It should be obvious that any proposal like Craig's for meeting our present demand for elimination of transcendental terms will be judged in various ways in accordance with varying approval of its tools and subsidiary concepts. Moreover, such variation may be independent of the question of modified empiricism as such. In particular, a dissatisfaction with systems containing infinite, effectively specified sets of postulates may or may not be justified, but is at any rate independent of modified empiricism as we have formulated it. Further, though the relevant notions of heuristic fruitfulness and suggestiveness, simplicity, and practicality are not very precise, suppose it granted that Craig's functionally equivalent system is indeed inferior to its counterpart in all these respects. This is irrelevant to our modified empiricism. If Craig's replacing system renders such empiricism possible, this represents an intellectual gain no worse for the fact that the system is unwieldy and not likely to be used by the practicing scientist. The case is analogous to ordinary definition, where we try to minimize the complexity of our primitive basis at the cost of replacing short and handy definienda by cumbersome definientia in terms of a simple few primitives. Obviously, no one intends these definientia to be used in practice in place of their definienda, but neither does anyone seriously maintain that their formulation therefore represents less of an intellectual gain.

One may however, with Goodman,[16] suggest the infinity of postulates in Craig's replacing system not as representing a practical difficulty, but rather as indicating that the deductive character of the original system is not sufficiently reflected by its replacement. That is to say, if transcendental theories serve to enable finite postulation, no replacement is equivalent *deductively* in every relevant sense if it fails to serve thus also, even though it does accurately reflect the whole class of relevant postulates-or-theorems (assertions) of the original. If specific empiricist programs are to be interpreted in accord with this point of view, then, even granted Craig's result, they are not proven generally achievable, and continue to represent non-trivial problems in individual cases. It seems to me, however, that if we take these programs as requiring simply the reflection of non-transcendental assertions into replacing systems without transcendental terms, then we do not distort traditional notions of empiricism, and we have to acknowledge that Craig's result does the trick; the further cited prob-

---

[15] C. G. Hempel, "Implications of Carnap's Work for the Philosophy of Science," to appear in the forthcoming Carnap volume of the *Library of Living Philosophers*.
[16] This point was made by Goodman in correspondence with the present writer.

lems remain but they are independent of empiricism as above formulated.

Professor Hempel's second reason against the scientific use of Craig's method is that "The application of scientific theories in the prediction and explanation of empirical findings involves not only deductive inference, that is, the exploitation of whatever deductive connections the theory establishes among statements representing potential empirical data, but it also requires procedures of an inductive character, and some of these would become impossible if the theoretical terms were avoided." He illustrates in terms of the following four sentences, where 'magnet' is taken to be a theoretical, that is, non-observational term:

> **(5.1)** The parts obtained by breaking a rod-shaped magnet in two are again magnets.
>
> **(5.2)** If x is a magnet, then whenever a small piece y of iron filing is brought into contact with x, then y clings to x. In symbols:
>
> $$Mx \supset (y) (Fxy \supset Cxy).$$
>
> **(5.3)** Objects b and c were obtained by breaking object a in two, and a was a magnet and rod-shaped.
>
> **(5.4)** If d is a piece of iron filing that is brought into contact with b, then d will cling to b.

Now, says Hempel, given (5.3) (and assuming (5.2)), we are able to deduce, with the help of (5.1), such sentences as (5.4). But (5.3) is non-observational, containing 'Ma', itself not deducible from observational sentences via (5.2) which states only a necessary, but not a sufficient condition for it. Thus, if (5.4) is to be connected by our theory here with other observational sentences, an *inductive* step is necessary, leading to (5.3), i.e., to 'Ma' specifically, from observational sentences. For example, 'Ma' might be *inductively* based on a number of instances of '$Fay \supset Cay$,' assuming that we have no instance of '$Fay \cdot \sim Cay$'. This is so, since such instances confirm '$(y) (Fay \supset Cay)$', which, by (5.2), partially supports 'Ma'. Thus, our hypothesis (5.1) takes us, in virtue of (5.2), from some observational sentences, that is, instances of '$Fay \supset Cay$', to observational sentences such as (5.4), but the transition requires certain inductive steps along the way. But though Craig's functionally equivalent system retains all the deductive connections among observational sentences of the original system, it does not, in general, retain the inductive connections among such sentences. Hempel concludes, "the transition, by means of the theory, from strictly observational to strictly observational sentences usually requires inductive steps, namely the transition from some set of observational sentences to some non-observational sentence which they support inductively, and which in turn can serve as a premise in the strictly deductive application of the given theory."

With respect to this argument, we might question by what theory of confirmation '(y) $(Fay \supset Cay)$' supports 'Ma'; it surely is not Hempel's satisfaction criterion of confirmation.[17] But this is irrelevant to the important point brought out by Hempel's argument; namely, that since functionally equivalent systems (of Craig's type) are not logically equivalent to their originals, they need not (on *any* likely view of confirmation) sustain the same confirmation relations as these originals, even among purely observational sentences. And this despite the fact that they do preserve the same deductive relations among such sentences by retaining all original theorems couched in purely observational terms. Thus, if we do not attempt an observational reduction of the *whole* of our theoretical discourse in given scientific domains via definition and translation or syntactic construction, but aim merely to isolate the *observational part* of such discourse, we must be careful to construe this part adequately, that is, as comprising not only a deductive network but also a wider confirmational range. Specific empiricist programs would then seem to be not achievable generally by means of Craig's result, in the light of Hempel's argument. One further line of attack might be to clarify the inductive relation sufficiently to enable agreement on just which sentences confirm which, relative to a given theoretical system, and then to strive for an independent, observational specification of such confirmation-pairs, to supplement the appropriate Craigian equivalent.[18] A second approach would be to try to meet each specific case by strengthening the replacing functional equivalent with hypotheses designed to yield just those inductive relations borne by the original, in which we are particularly interested.

In the case of Hempel's above example, for instance, the inductive relation in question is between instances of '$Fay \supset Cay$' plus the observational parts of (5.3), (objects b and c were obtained by breaking object a in two [$Bbca$] and a was rod-shaped [Ra]) and (5.4). Such an inductive relation might also be expressed without theoretical terms by the following statement, which, owing to the universal quantification, can at best establish only an inductive relationship between its instances and the observational sentence derived:

[17] "A Purely Syntactical Definition of Confirmation," *Journal of Symbolic Logic*, VIII (1943) 122–43. See also C. G. Hempel, "Studies in the Logic of Confirmation," *Mind*, N. S. LIV (1945), 1–26 and 97–121, especially 107 ff.

[18] It would also be desirable to clarify the notion of induction sufficiently to enable critical evaluation of the assumption that where the Craigian equivalent fails to reflect the confirmation relationships of the original theoretical system, it is itself confirmed to a lesser degree by the available evidence on which that original rests. Perhaps even more interesting would be the examination of the analogous assumption for the *strengthened* functional equivalent discussed immediately below.

$$(x) \{[(y) (Fxy \supset Cxy) \cdot Rx] \supset (z) [(\exists w) (Bzwx) \supset (u) (Fzu \supset Czu)]\} [19]$$

19. Summary and Conclusion

If our journey has yielded no single, easy solution as a climax, its difficulty has nevertheless earned for us the right to stop and get our bearings. For we have traveled a long way from the conception of empiricism as a shiny, new philosophical doctrine for weeding out obscurantism and cutting down nonsense wherever they crop up. We have, furthermore, seen that even if we take empiricism as the proposal of a general meaning-criterion in terms of translatability into a chosen artificial language, we run into trouble. We have thus come to restrict the empiricist's job to providing merely an adequate sufficient condition of significance on an observational basis, in the form of an observational system capable of housing our scientific beliefs.

Even this restricted task has, however, turned out to have quite difficult obstacles before it. While the inclusion of needed disposition terms seemed to us not as formidable a problem as hitherto thought, we found theoretical terms to be generally resistant to straightforward empiricist interpretation. Considering this difficulty in the light of a number of recent approaches to philosophy of science, we found that the pragmatic rejection of our restricted empiricism does not constitute a refutation, while instrumentalism's easy solution fulfills such empiricism in only the most trivial sense. Taking empiricism's task as the provision of an appropriate modification of scientific discourse itself rather than simply of our notions of belief, we found the possibility of syntactic reinterpretation promising, though less intuitively satisfying then a direct reinterpretation of the object-language of science proper. Our final examination of Craig's method for eliminating theoretical terms wholly from such language while preserving its observational segment intact led to the conclusion that this method is, in itself, incapable of achieving the goal of our restricted empiricism.

It appears, in sum, that even a modest empiricism is presently a hope for clarification and a challenge to constructive investigation rather than a well-grounded doctrine, unless we construe it in a quite trivial way. Empiricists are perhaps best thought of as those who share the hope and accept the challenge—who refuse to take difficulty as a valid reason either for satisfaction with the obscure or for abandonment of effort.

Harvard University

[19] This formula is due to Professor Hempel, who suggested it to me as an improvement over my original version.

J. J. C. SMART

The Reality of
Theoretical Entities

A theoretical physicist normally has a certain characteristic picture of
the world. For him the world consists of unimaginably large numbers
of electrons, protons, neutrons, photons, and so forth. These ultimate
particles are for him the stuff of the world. We have found out about
these particles by somewhat indirect methods, he would say, but nevertheless to assert that the world consists of such things differs only in
degree from saying that a certain powder consists of small crystals,
visible only under the microscope.[1] No one doubts that such crystals
exist and are real in the same sense as that in which the powder exists
and is real. Why not electrons too? On this view electrons are as real
as chairs, and things consist of electrons, protons, and so forth, in the
way that a wall consists of bricks.

Under the strain of certain conceptual difficulties arising out of
relativity and quantum theory some physicists have adopted a quite
different point of view. According to this doctrine the fundamental
particles are no longer regarded as actual real things in the way that
(perhaps) chairs, galvanometers, and crystals are, but are thought of
rather as logical constructions out of our observations, fictions which
enable us to state compendiously numerous apparently heterogeneous
laws of the interrelations of observations. In this they concur with
older philosophical scientists who advocated a similar doctrine under
the strain of certain sceptical arguments of a general philosophical
nature, hinging on concepts such as *explanation, necessity* and *real*.
Often the two sorts of conceptual difficulty are intermingled, and frequently the arguments of a scientist who advocates that electrons are
only logical constructions out of observations hinge much more on

Reprinted by kind permission of the author and publishers from *The Aus-
tralasian Journal of Philosophy*, Vol. 34 (1956), 1–12.

[1] *Cf.* Max Born, 'Physical Reality', *Philosophical Quarterly*, Vol. 3 (1953), 139–49,
especially 141.

general concepts like 'real', 'explanation', and 'necessity' than on specific unclarities about electrons, simultaneity, or indeterminacy. Frequently the view that electrons are logical constructions out of observations with cathode ray tubes, galvanometers, and so on, is combined with and draws support from a phenomenalist view that the cathode ray tubes and galvanometers themselves are only logical constructions out of sense-data.[2] On this view theoretical entities are constructions out of constructions, and so do not differ importantly from our familiar tables and chairs, which are themselves only constructions. On this extreme phenomenalist form of the 'logical construction' doctrine theoretical entities are once more assimilated to everyday objects, but at the price of denying bedrock reality to both.

I should make it clear that in this exposition I am not using the term 'logical construction' in its strictest sense. In discussing the view that theoretical entities are logical constructions out of observations with cathode ray tubes, galvanometers, and so forth, I am not discussing the view that statements about theoretical entities can be translated into statements about observations, for this view has been conclusively refuted by F. P. Ramsey and R. B. Braithwaite.[3] I am considering the allied view that somehow or other talk about electrons is just a matter of talking abstractly about observations, in a way similar to, though different in some respects from, the way that talk about nations is a matter of talk about people. (Even though nation statements cannot be translated into person statements.)

In this dispute most professional philosophers seem at present to be more or less on the side of the 'logical construction' doctrine (though not, usually, in its extreme phenomenalist form). Thus Mayo says 'If theoretical concepts are only to be explained by referring to the particular theory in which they have a place, and if the only justification for accepting a particular theory as established is that it enables us to explain known facts and successfully predict new ones, then it follows that we gain nothing by either attaching or withholding the epithet "real" '.[4] To say that electrons are real can be to say no more than that a certain concept occurs in an established theory. This is to give electrons a very thin sort of reality: it puts them on a level with lines of force, accelerations, and even geometrical points, all of which are concepts which occur within physical theory. Can we do no more than this? Even if the physicist who says that electrons are as real as microscopic crystals of a powder should be wrong, is he not making a real distinction? Such a man would by no means have an

[2] *Cf.* H. Margenau, *The Nature of Physical Reality*, 1950.

[3] Ramsey, 'Theories', in *The Foundations of Mathematics*, Braithwaite, *Scientific Explanation*, chapter 3.

[4] B. Mayo, 'The Existence of Theoretical Entities', Penguin *Science News*, 32, 17 (1954).

inclination to say that a line of force, still less a geometrical point, is on a par with a small crystal of the powder. Toulmin, in what is perhaps the best and clearest short discussion of the problem of the reality of theoretical entities,[5] after adopting a position very like that taken up in Mayo's later article, does do something to restore the distinction in question. 'Do atoms, genes, electrons, fields, etc. exist?' has, he suggests, the force of 'is there anything to show for them?' 'To a working physicist, the question "Do neutrinos exist?" acts as an invitation to "produce a neutrino", preferably by making it *visible*.' [6] Obviously, however, one cannot literally *see* an electron or a neutrino, and so 'our problem is accordingly complicated by the need to decide what is to count as "producing" a neutrino, a field or a gene. It is not obvious what sorts of things ought to count: certain things are, however, generally regarded as acceptable—for instance, cloud-chamber pictures of X-ray tracks, electron microscope photographs or, as a second-best, audible clicks from a Geiger counter." The trouble about this is the apparent arbitrariness of the collection of procedures which count as 'producing' an entity or 'something to show' for it. One has an inclination to say that lines of force are not entities in the way that electrons are, but there can plausibly be said to be 'something to show' for them: indeed magnetic lines of force can be in a sense shown very vividly by scattering iron filings near the poles of a magnet. And on the other hand even so vivid a thing as seeing a large molecule in the electron microscope can be thought of sceptically. What is seen, it can be said, are only dark patches on a cathode ray screen, meaningless without the theory of the electron microscope which enables us to interpret them, and the large molecule is just a fiction which may enable us to conflate these observations with others of a different sort.

To put my cards on the table, let me say that there is a sense in which electrons, neutrinos, photons, and so on are real things (whose existence is no more abstract than that of the small crystals of our powder) as opposed to lines of force, points of space, or components of momentum, which can plausibly be dismissed as constructions or abstractions. (Like the 'factors' of factor analysis in psychology, which no reputable psychologist thinks of as actual things.) Of course, on some theoretical pictures, an electron might be described as a singularity in a field of force, but then this field, whether Descartes' extension or the modern space-time, is thought of intuitively as an extensive substance, whose warpings constitute the field. This field picture is just one way of describing what is described by the discrete atomistic picture, and which picture is adopted does not affect my argument. Let

[5] *The Philosophy of Science*, 1953, pp. 134–39.
[6] *Philosophy of Science*, p. 136.

us see how far my crude statement of a naïvely realistic picture can be discredited by philosophical analysis. One can already hear one's philosophical friends saying impatiently: 'Of course this is a typical philosophical puzzle. When in philosophy we ask whether we should say "yes" or "no" the answer is always "yes and no" or "neither and both". Of course there are features of the logic of "electron" that pull us in the direction of classifying electrons with bricks. And there are other features that pull us in the direction of classifying them with factors or geometrical points. When all these features have been described, then say what you like.' But is the answer to the question necessarily 'yes and no'? Might it not be at any rate very much more 'yes' than 'no' or very much more 'no' than 'yes'?

An initial difficulty that presents itself is this: that it is not easy to think of situations in which the question 'do electrons exist?' would naturally arise. A question like 'do unicorns exist?' is a clear question because the concept of a unicorn is a common-sense one which can be easily enough explained in terms of other concepts: a unicorn is a quadruped with one central horn. Whether this concept applies to anything or not is something we can find out in a straightforward empirical way, say by searching the southern Sahara desert. In order to learn the use of the word 'electron', however, we have to learn what an electric charge is, what a cathode and an anode are, about J. J. Thomson's experiments, Millikan's experiments, and so on. Fully to understand the word as it is now used we must furthermore understand the part 'electron' plays in theory, know something of the structure of the atom, and about the successively more refined developments of quantum mechanics that have grown up. Only after all this will we know how to use the word 'electron' as physicists use it. Learning the meaning of the word is something which comes to something little short of learning physics. When we have learnt a modicum of physics the question 'do electrons exist?' no longer arises. Before we reach this stage this question has no very clear meaning. Nevertheless at an early stage it has *some* meaning. We can imagine Thomson and Millikan summing up the results of their experiments by saying 'electrons exist': that is, 'a smallest discrete electric charge exists, which appears to be associated with a particle'. Even now, though the question 'do electrons exist?' does not arise, that is not, I think, to say that it has no meaning. The question 'does anything exist?' does not arise, but surely it has meaning and the answer to it is 'yes'. We can imagine present-day physical concepts being overthrown by new ideas, and its being denied that there are any quanta of electric charge. (We can *imagine* this: I do not say that we can believe this.) In such a case 'do electrons exist?' could be answered by 'no'. If a future historian of science made sufficiently arduous researches he could come to learn the rules according to which the word 'electron' was once used: the fact that on our supposition the theory of electrons and these rules with it would now be on the scrap heap does not affect the matter. And if asked 'do electrons exist?' he could answer 'no'. But admittedly the question 'do electrons exist?' would not then naturally arise. (Nor does it arise now when the theory is taken for granted.)

Since the question 'do electrons exist?' does not naturally arise it follows that if anyone did ask it he would be taken most naturally to be asking a question about the logical status of the concept 'electron'. There are two ways in which a question of this sort might be asked. In the first place we could consider the question as a quasi-empirical one. That is, we might ask whether the experimental facts are such that they push us in the direction of thinking of electrons as actual particles rather than as avowed fictions like lines of force or geometrical points. In the second place, when this quasi-empirical question has been answered in the affirmative there remains a purely logical question: what is it to think of electrons as actual things, not as avowed fictions? Are not these actual particles themselves in a subtle way perhaps no more than theoretical constructions? That is, the question 'are electrons real?' can be construed in a wholly non-empirical way as a question not about electrons but about the use of the *word* 'electron'. In order to substantiate this last suggestion and in order to see just what it amounts to, it will be necessary to pay some attention to the logic of the word 'real'.

What do we mean when we say that elephants are real? At first sight this is to say the same as that elephants exist, or that there are elephants. Nevertheless if we look closer we can see that '. . . is real' normally does a somewhat different job from 'exists' or 'there is a . . .'. There is no harm in saying that there is a prime number between 15 and 20 and that such a prime number exists, but to say that a prime number between 15 and 20 is real would be very odd. There is no harm in saying that there is a mad king in one of Shakespeare's plays, but most of us would deny that such a person was real. Closely connected with this (grammatically) predicative use of 'real', in which we rebut the suggestion that the thing in question is an abstraction, or a fiction, or imaginary, there is a purely adjectival use in which we say that something is 'a real such-and-such'. When we say that a diamond is a real one we rebut the suggestion that it is an imitation one.[7] To say that the conquest of Everest was a real achievement is to rebut the suggestion that it was a meretricious one. Notice that if we doubt whether a diamond is a real one or whether a piece of material is real silk or whether an achievement is a real achievement we do not doubt whether the thing in question exists (or whether the event in question

---

[7] *Cf.* Mayo, 'The Existence of Theoretical Entities'.

actually happened). Furthermore, and this is more important, there is no category difference between the real thing and the imitation, counterfeit, or meretricious one. The real achievement and the meretricious one are both events, the real silk and the artificial silk are both stuffs, and the real diamond and the imitation diamond are both things. On the other hand, when we use the word 'real' in the predicative way, as in 'leprechauns are not real: they are imaginary', we are making a category distinction. A real person and an imaginary person are not two examples of the same sort of thing in the way that a real diamond and an imitation one are. 'Leprechauns are not real' is a disguised statement about the way the word 'leprechaun' normally works: that is, only very naïve people play the same language game with the word 'leprechaun' as they do with 'lions' and 'persons'. There are thus two logically different ways in which we use 'real'. We use it to make a distinction of fact (real versus imitation, etc.) and we use it to make one of a number of category distinctions (real versus imaginary, fictional, abstract, etc.). (There is not just *one* category distinction that is made: for example the distinction real (actual) versus imaginary is quite different from that of real (concrete) versus abstract.) I have tried to relate these two sorts of uses of 'real' to differences in grammatical usage: the predicative, '. . . is real', and the purely adjectival, '. . . is a real . . .'. However, grammar is not an infallible guide. We often say leprechauns are not real things' or 'leprechauns are not real creatures' instead of 'leprechauns are not real', and we can say 'that diamond is not real', meaning 'that stone is not a real diamond'. Note, however, that in the former case we tend to finish the sentence with a category, or near-category, word like 'thing' or 'creature'. To use a non-category word like 'men' and say 'leprechauns are not real persons' would suggest that they were in some way artificial or imitation men, which is not what we want to say at all.

Mayo (in 'The Existence of Theoretical Entities') says: 'Anyone who wishes to maintain that electrons exist or are real must be prepared to ask himself what possible sorts of defectiveness or lack of qualification he is concerned to deny.' We may question the implied suggestion that to say 'electrons exist' is much the same as to say 'electrons are real', and so let us for clarity replace 'exist or are real' in the above statement simply by 'are real'. Then, suggests Mayo, if we say that electrons are real we are denying some sort of lack of qualification or defectiveness. If we cannot say what sort of lack of qualification or defectiveness is in question then our assertion lacks a clear meaning. On the other hand we have seen that though 'electrons are real' is closely connected with 'electrons are real such-and-such's' there is a difference between the two statements. If a man said 'electrons are not real particles' we might understand a certain sort of defectiveness to be in his mind: say, that electrons are wave packets rather than particles in the traditional sense. But if a man said 'electrons are not real' he could be described *either* as not asserting a defectiveness at all *or* as asserting a logically queer kind of defectiveness. To take a parallel example. If a man said that numbers are not real he would perhaps mean that numbers are not concrete things but abstractions. Some philosophers like Nelson Goodman might object to numbers on this score: a universe containing such dubious entities could in a sense be described as a universe containing defective elements. Goodman could in a sense be described as a man who regarded abstractions as defective, not fit for his world as concrete objects are. On the other hand we can also see that the above way of putting the matter is highly misleading. If there are (in a sense) no such things as abstractions how can we say that there are such things but that they are defective? It is clear that what Goodman would regard as defective is not the abstractions themselves but certain words (words which purport to name such entities). What Goodman wants is a certain sort of language, a nominalistic one. It is not numbers that are defective in his view: it is numerals (at least as used in a certain way).

In the same sort of way, if we say that electrons are real or not real we are not (to use Mayo's words again) denying or asserting some defectiveness or lack of qualification (as we would be if we said that they were or were not real particles), or, to put it differently, we would be denying or asserting defectiveness or lack of qualification in a queer sort of way. The sort of defectiveness or lack of qualification we would have in mind would be 'being an abstraction' or 'being a theoretical fiction' or 'being a logical construction', as opposed to being a concrete entity like a chair, a microscopic crystal, or a microbe. If we spoke in this way we should be speaking inaccurately or misleadingly: it would be better to say that what we are really accusing of defectiveness is the word 'electron', that is, we should be saying that the word 'electron' belonged to a certain logical category. And if we have a preference for a language containing certain logical categories rather than others, or if we prefer to talk about concrete entities rather than abstractions, to say 'electrons are not real' would be to complain that though the naïve scientist supposes that 'electron' behaves like 'microbe' it really behaves more like, say, 'root mean square velocity'. That is, our scientific language is accused of being less concrete than we had hoped.

Let us go back to considering the fine powder whose crystals are visible only under the microscope. We feel confident enough (in a way we do not feel confident about electrons) that the crystals are concrete entities just as bricks and houses are. What is the basis of our confidence?

Suppose a person looks through a very low power microscope. He sees a lump of sugar, say, looking twice as large. He does not need to know the theory of the microscope to know that what he sees is a lump of sugar. It looks just like a lump of sugar, only bigger. Even a savage, who knew nothing of optics, would tend to say that what he saw was a lump of sugar. It would be even more likely that he would say that what he saw through the lenses was a lump of sugar if he also had a look with the naked eye at the lump of sugar on the tray of the microscope. Compare the way children automatically say 'car' when they see a picture of a car: no doubt they do this even before they have any conception of what a picture is. (Certainly they may before they learn the use of the word 'picture'.)

In this way we get to think that a microscope shows us what things would look like to us if they were many times bigger (or we many times smaller). We thus think of the minute crystals of the powder as only contingently invisible to our naked vision, and we think of it as only contingent that they do not seem as substantial to our naked vision as bricks and motor cars now do. This way of thought is naturally strengthened by what we know of optics.

When we pass from what we see with ordinary microscopes to what we see with the ultra-microscope or the electron microscope a different situation presents itself. In such a case we could not expect our savage to interpret what he sees: here knowledge of optics or electron optics is essential. Nevertheless, given this knowledge, we can see that it is still a contingent fact that we cannot see with our naked eyes what we see with the microscope. For example, if there were as many electron beams through space as there now are photon beams, and if we had ultra-microscopic eyes somehow constructed on the electron microscope principle, we might be as much at home in the world of individual living cells as we now are in the world of dogs and cows. (I neglect here certain difficulties: one is that if there were as many electrons flying around as there are photons now it might be incompatible with what we know of biophysics that our living cells should have the structure they now have.)

It now looks no more than a contingent fact that we cannot see even such recherché entities as bacteria, viruses, cells, and even large molecules. What if we go further and say that it is just a contingent fact that we cannot see electrons, protons, mesons, photons, and so forth? Here, however, theoretical considerations come in. In any ordinary sense of 'see' what we see must have a definite position and shape, and our seeing of it must leave it substantially unaffected. These conditions could not hold with fundamental particles. Since 'electron' and so forth get their meaning from the part they play in theory, we cannot say that it is only a contingent fact that those conditions hold which make it physically impossible for us to see them. In the case of a photon it is particularly obvious that however microscopic we were we could never see a photon. Photons explain seeing: they therefore cannot themselves be seen. For would one see a photon by means of a further photon? Furthermore, if photons could be seen we could never see anything else: they would get in the way of whatever else we wanted to see.[8] The whole conception of 'seeing a photon' is self-contradictory. 'Photon' gets its meaning from the part the word plays in prevailing theory, and this theory in turn rules out the possibility of photons being seen by anything, however microscopic.

There thus seem to be two reasons, and two only, for putting the names of the fundamental particles in a different category from 'brick', 'microscopic crystal', and 'cell'. One is that there are theoretical reasons why they could never be perceived: it is therefore not a contingent fact that we do not talk about them in a language at all analogous to our ordinary everyday language of perceptible objects. The second is that if the theories in which their names play a part are given up or radically modified (as they may well be) we shall give up even talking of electrons, protons, and so forth. This, however, is not as solid an objection as seems at first sight. There is no reason to suppose that 'electron' will ever suffer the fate of 'phlogiston'. Whatever it is that we now describe (possibly misdescribe) in our present theoretical language will still have to be described (or misdescribed) in the new language. After all, when in the 1920's people stopped talking about particles in the sense of the Bohr atom and started talking about wave packets, 'wave packet' still tried to describe what 'particle' did before. The situation is not all that different from the sea serpents which later mariners describe as schools of porpoises: there is something described or misdescribed by both the old and the modern sailors.

It might be objected here that the new description might involve a category difference. But this might be so even with the sea-serpent: 'sea-serpent' might misdescribe not a group of porpoises but a wave, and a wave is of a different category from a serpent. Nevertheless the mariners who said they saw a serpent and those who said they saw a wave might, in a sense, be said to be 'describing the same thing'. Category distinctions are in any case by no means hard and fast: if you try hard enough you can exhibit any word as in a different category from any other word.'[9]

The situation seems to be this: there are things which we cannot describe in detail but about which we can already say quite a lot. We cannot see them and there are theoretical reasons why we could never in principle see them. And that is about all there is to it.

I should like, furthermore, to defend the no doubt naïve view of

---

[8] I think that I once heard this point made by Dingle.

[9] See my 'Note on Categories', *British Journal for the Philosophy of Science,* vol. 4, 227–28.

most physicists that modern theoretical science gives us a truer picture of the world than does the language of ordinary common sense. This is not to say, of course, that the propositions of everyday life are not true. Of course they are true. That is, I want to specify a sense in which physical theory gives us 'a truer picture of' the world than does common sense, which does not entail that theoretical statements are more frequently true than common-sense ones. Now there is one very clear sense in which we can say that physical theory gives us a truer picture of the world than common sense does. This is that many of the concepts of common life are anthropocentric. It is colour concepts that I shall single out for consideration, but there are of course others. (For example 'nauseating', 'sickly', 'dazzling', 'sweet'.) It is a well known fact that various mixtures of wavelengths of light will occasion the same discriminatory response from a normal person as pure light of a certain single wavelength will. Just what mixtures of light are equivalent to one another in this respect is a rather arbitrary matter, depending on the peculiarly complicated structure of the eye and nervous system. We have by no means got 'spectroscopic eyes'.[10] It would be much easier to construct an instrument which would distinguish between any two different distributions of intensity of light among wavelengths than to construct an instrument which makes the particular discriminations, and lack of discriminations, that the human eye makes. (I have no doubt that the latter task is in practice impossible.) This shows the arbitrary and anthropocentric nature of our colour concepts. There is no reason why Martians should have the colour concepts we have. We would, however, expect them to have the same sort of physics, though it might be more or less advanced than ours. 'A truer picture', then, means 'less (or not at all) anthropocentric'. One way in which scientific language is less anthropocentric than ordinary language is that our descriptions of the world are not confined to macroscopic objects. And with this goes greater detail of description and proportionately great simplicity of explanation.

I have been trying to cut down to life size the philosophical objections so commonly brought against regarding theoretical entities as concrete things. It is true that many physicists have tended to adopt a sceptical attitude to the concrete reality of theoretical entities under the stress of the difficulty of reconciling a more or less realistic view with the present state of atomic physics. It is beyond my province to deal with such objections: I should merely like to state that I have on my side one very eminent physicist (Max Born) whose arguments[11] against such objections seem to me eminently sensible. It is also evi-

---

[10] Cf. Pope, *Essay on Man*: 'Why has not man a microscopic eye? For this plain reason: man is not a fly.'

[11] Max Born, 'Physical Reality'.

dent that most physicists who base their arguments on the modern state of physics often give expression to phenomenalist strains of thought, and it is open to question how many of their difficulties stem from quandaries about recondite physical concepts, and how many, after all, are just the familiar philosophical parrot cries.

The naïve physicist thinks that his science forces us to see the world differently and more truly. I have tried to defend him in this view. The modern tendency in philosophy is to be opposed to phenomenalism about tables and chairs but to be phenomenalist about electrons and protons. Philosophers thus tend to give an ontological priority to everyday concepts and they come to have 'a sharply . . . definite view of the world: a world of solid and manageable objects, without hidden recesses, each visibly functioning in its own appropriate pattern'.[12] They are wedded to the concepts of common sense every bit as much as the mediæval pre-Galilean philosophers were, but they are less open about it.

Adelaide University

---

[12] The quotation is from the last paragraph of S. Hampshire's critical notice of Ryle's *Concept of Mind* (*Mind*, vol. 59, 1950). I do not think, however, that a phenomenalist doctrine about theoretical entities is in fact entailed by Ryle's book.

GROVER MAXWELL

Theories, Frameworks, and Ontology

Carnap's classic essay, "Empiricism, Semantics, and Ontology" (Reprinted in *Meaning and Necessity,* 2nd ed., Chicago Univ. Press, 1955) —hereafter: *ESO*) contains, I believe, the basis for the definitive solution of all significant ontological problems. The modest aim of this brief paper (the responsibility for which is, of course, entirely my own) is to add some detail to a part of the Carnap approach and to apply it specifically to the problem of the ontological status of theoretical entities.

Carnap's central contention seems to be that in order to talk about any kinds of entities at all and thus, *a fortiori,* to talk of their "reality" or existence, it is first necessary to accept the "linguistic framework" which "introduces the entities." What does "acceptance" of such a framework involve? First, let us consider some of the conditions which *any* linguistic framework (i.e. any language) that talks descriptively about the world should fulfill. There will be, of course, a set of L-formation and L-transformation rules (the "purely syntactical" rules) and the corresponding set of L-truths which they generate.* For simplicity's sake, let us assume that no *explicitly* defined terms are introduced. Although this be assumed, I shall contend (again following Carnap)[1] that any such framework should contain certain sentences which are *analytic* (in a broad sense) and, thus, "factually empty" but

---

Reprinted by kind permission of the author and publishers from *Philosophy of Science,* Vol. 29 (1962), 132–38.

* For an explanation of the terms "*L*-formation" and "*L*-transformation" see Carnap, pp. 33–35.

[1] See Carnap, "Meaning Postulates", *Philos. Studies,* 3:65–73 (1952) and "Beobachtungssprache and theoretische Sprache", *Dialectica,* 12:236–48 (1957), and also my "Meaning Postulates in Scientific Theories", (hereafter: MPST) in H. Feigl and G. Maxwell (eds.) *Current Issues in the Philosophy of Science* (hereafter: CIPS) New York: Holt, Rinehart, and Winston, Inc., 1961, pp. 169–83.

which are *not* L-true. Let us call such sentences "A-true." A set of such sentences might include, 'All physical objects are extended', 'No physical object can be in two places at once', 'All chairs are physical objects', 'No surface can be simultaneously red and green all over', 'If event A is earlier than event B, then B is not earlier than A', as well as 'All electrons are electrically charged', 'The one-way speed of light is independent of direction' [2] and many others. (For convenience I have used ordinary English, but the same considerations apply to any *constructed* language which is given empirical application.) The set of all L-independent A-true sentences which contain a given term may be said to *implicitly define* the term, that is the *non-ostensive* meaning of the term is completely given by such a set. Consider, for example, a so-called "theoretical term." The theoretical postulates which contain the term (including the so-called "correspondence rules") will fall into two sets, one containing the A-true sentences which implicitly define the term (as well as being part of the implicit definition of the other terms they contain), and the other set containing contingent statements, which are subject to empirical confirmation or disconfirmation.[3] Furthermore, a descriptive framework must contain a set of sentences whose truth value is *quickly decidable* on a non-linguistic basis (i.e. non-inferentially decidable). These correspond, of course, to the so-called "singular observation statements." It is neither necessary nor desirable that such statements be incorrigible or indubitable or that a sharp distinction between observational and theoretical terms be drawn. Finally, for any such framework, there must be a set of rules or procedures whereby *universal statements* (observational *and* theoretical) and *singular theoretical* statements are confirmed and disconfirmed.

It is evident that the burden of giving the "essential nature" of a kind of entities and, *a fortiori,* that of introducing *new* kinds of entities falls upon the set of A-true sentences containing terms that designate the entities in question. Such sentences should fulfill the following requirements: (1) They must be consistent with the pre-existing A-true sentences of the framework or the latter must be modified so that the total set is consistent, (2) the conjunction of the total set of A-true

---

[2] For arguments that this particular sentence, as it is used in physical theory today, is "conventional" and, thus, factually empty and A-true, see A. Grünbaum, "Geometry, Chronometry, and Empiricism", in H. Feigl and G. Maxwell (eds.) *Minnesota Studies in the Philosophy of Science* (hereafter: MSPS), Volume III, Minneapolis: University of Minnesota Press, forthcoming.

[3] Here I depart considerably from Carnap's latest interpretation of theories, which is outlined in the *Dialectica* article (in "Meaning Postulates") and which employs only one A-postulate—a postulate which contains the entire scientific theory (plus correspondence rules) in question along with its Ramsey sentence. While this kind of reconstruction is viable and in many ways illuminating, I feel that where one is concerned the meanings of individual terms, a less global approach, such as the one I have just outlined, is preferable.

sentences must in fact *be* A-true, that is the rules introducing the new kind of entities must not have contingent consequences, and (3) the meaning given the new term(s) should fall within a vaguely specified interval of a "measure of richness," which I can only adumbrate here by means of a simple example. Suppose we purport to introduce a "new" kind of entities, glub-glubs, by means of the one A-true sentence, 'Something is a glub-glub *if and only if* it is a horse'. The meaning of 'glub-glub' here is too rich; we do not want to say that the introduction of a mere synonym introduces a new kind of entities. Suppose, on the other hand, glub-glubs are "introduced" by the *one* A-true sentence, 'Something is a glub-glub *if* it is a horse'. Here the meaning of 'glub-glub' is too *poor*. Although the reality (existence) of glub-glubs would follow from the existence of horses and although the class of glub-glubs is not identified with the class of horses, the question remains: What is the point of introducing the term 'glub-glub' here at all?

Now let us suppose that Mr. A and Mr. B have some framework, *L*, in common, that is that they can communicate by means of *L*. Mr. A says, "I purpose to enrich *L* by connecting terms referring to a new kind of entities, ϕs, to some of the terms, statements, and so forth of our framework by means of appropriate A-true sentences." Now, provided A's A-true sentences meet the requirements listed above, nothing but sheer perversity can prevent B from giving him a sympathetic, tentative hearing. Then if A takes the *additional* step of asserting the reality of the new kind of entities, *that is of asserting that ϕs exist*, he must produce a *contingently* true sentence (using the confirmation rules of *L* along with some true, quickly decidable statements) which entails the sentence, 'There are ϕs'. If he does so, then it seems to me that B must agree, "Yes, in your sense of 'ϕ', ϕs do exist;" however, he *may* want to add, "But why talk that way? Why not talk in the manner I am about to propose? It's simpler (or more convenient—or more beautiful)."

Let us consider a simple example. Mr. A proposes to describe surfaces which Mr. B (along with most of us) would call 'chartreuse' as being both yellow and green all over. "Furthermore," he argues, "an almost opaque solution of fluorescein presents what, in a sense (the *sense which I am proposing*), is a striking example of a surface which is both green and yellow all over, and such a solution of merthiolate is simultaneously both red and green." [4] Mr. A. also introduces the class term ("kind" word)[5] 'multicolored surface' to designate his "new kind of entities." Suppose that Mr. B demurs, insisting on locutions such as 'chartreuse', 'reddish-green', and so forth. Then the frameworks

[4] For interesting, related considerations concerning "multicolored" surfaces, see J. J. C. Smart, "Incompatible Colors," *Philos. Studies*, 10:39–42 (1959).
[5] Cf. Wilfrid Sellars, "Grammar and Existence: a Preface to Ontology," *Mind*, forthcoming, and Carnap [ESO].

of A and B would differ in that they would embody, respectively, slightly different meanings of words such as 'colored surface', 'red', 'green', etc.; for example, in B's framework, the sentence 'No surface can be simultaneously red and green all over' would be A-true, while in A's it would be *contingently* false. And, since in principle both A and B can, each in his respective framework, express and explain all the "facts" concerning *colored surfaces* (in either sense), "the facts" can never compel either to accept the other's framework[6]; and when A and B utter, respectively, 'There *are* multicolored surfaces' and 'There are *no* multicolored surfaces', they are *not* expressing any cognitive disagreement—simply because they are not using 'multicolored surface' in the same sense. Taken as "internal assertions" (Carnap [ESO]) in their respective frameworks, each statement is (contingently!) true. Construed as putative answers to an "external question" (*ibid.*), the sentences are *not* cognitive assertions; at most, they merely *evince* A's and B's attitudes toward their respective frameworks.

Let us now examine, in more detail, a question about which Carnap has been tantalizingly terse: What factors are relevant for the "pragmatic" justification (vindication) of a framework?[7] Or: What is it for one framework to be *more useful* than another? I believe that there are two problems here which have often been conflated. (We shall see, however, that, indeed, they are *not* completely extricable from each other). The first is: What requirement must *any* descriptive framework worthy of the name fulfill? The second is: Given that two or more frameworks fulfill this requirement, what additional factors are relevant for our choice of one framework over the others?

The answers to the much easier second question are, I believe: simplicity; ease of comprehension, communication, and computations and other inferential manipulations; and also aesthetic considerations (including, even, personal idio-syncratic tastes, provided we can persuade other relevant language users to accept our frameworks). I shall call these factors "purely pragmatic" or *useful* in the weak sense.[8]

As a crude stab at the first question we might say that a framework must provide the linguistic apparatus for "saying all that we want to

[6] However, if examples similar to fluorescein and merthiolate and others even more striking, which are now beyond the limits of our imagination, began to turn up more and more frequently, "the facts" might "gently persuade" (see my MPST) B to adopt A's framework. For such a way of talking might turn out to be a simpler—or aesthetically more pleasing—or a more convenient (though no more "logically adequate") means of communication.

[7] For an excellent discussion of a number of the issues involved here see H. Feigl, *"De principiis non disputandum?"* in M. Black (ed.) *Philosophical Analysis,* Ithaca: Cornell University Press, 1950.

[8] Although, of course, this is an extremely important sense. A framework which was so cumbersome that comprehension, communication, or computation within it was virtually impossible would be virtually *useless* in any sense of the word.

say" or for reporting, explaining, and predicting all of "the facts." Neither of these answers is very helpful; for, as Carnap says [ESO], some of us might "want" to refrain from speaking altogether, and just what "the facts" *are* is, of course, part of what is at issue (i.e. what the world consists of, what exists, what "reality" is "really" like, etc.). Our experiential access to the facts, whatever these may be, is, of course, via the *quickly decidable* statements of our framework. If we are imbued with a reasonable amount of curiosity and a desire to increase our store of knowledge, then both our individual acts of tokening such statements and our tendency to increase their number and their variety (by, say, increasing the number of primitives employed in them) are "forced upon us" in a manner more easily "felt for" than described.[9] Now while we are clearly about as close to "rock bottom" here as we shall ever get, it would be a mistake to comfort ourselves with labels such as "the Given," "direct experience," "directly observed," and so forth and to suppose that the "observational" portion of a framework is immune to rational and theoretical appraisal and modification. We have already seen how this might be done even for the "primitive observation" predicates 'red' and 'green'. *Nevertheless,* since the quickly decidable statements are in some sense a sort of starting point or "jumping-off place," the fact that a framework allows us to express a greater number and a larger variety of "observational facts" and—*and this is crucial—to explain these facts* is a good indication that it approaches more closely the requirement mentioned in the first question[10] —a requirement I have only been able to crudely formulate. (I shall call this requirement the *condition of adequacy* of a framework or *usefulness in the strong sense*—although it seems a little misleading to use so weak a word as 'usefulness' here at all.) *Explanation* is important, partly for its own sake, partly for its role in prediction, but most of all because experience *and* analysis show that the *facts* about the ("theoretical") entities which we invoke in our explanations comprise an indispensable realm of the totality of "facts about the world." This

---

[9] The pseudo-epistemological task of accounting for our ability to use the quickly decidable statements to make "factual reports" can, indeed, be accomplished—or, rather, it could be accomplished if we had adequate scientific theories of perception, brain processes, "the tokening mechanism," etc. But such an account would *have* to employ a framework including, of course, *some* quickly decidable statements. For this and for reasons given in the text, epistemology must always contain, essentially, an element of circularity (Carnap, I believe, has also made this point). I use the term 'pseudo-' above not as an epithet but to emphasize that such an "epistemological" account cannot be given using only "purely logical," or "purely conceptual," or "purely metaphysical" considerations. For helpful considerations concerning these problems see Wilfrid Sellars, "Some Reflections on Language Games," *Philosophy of Science* 21:204–28 (1954) and "Empiricism and the Philosophy of the Mind," in MSPS, Vol. I.

[10] Cf. Wilfrid Sellars, ". . . Language Games," p. 67.

can be argued for on many grounds,[11] perhaps the most obvious of which is the fact that, as our theories and instruments for observation are developed and improved, what were once "theoretical" entities become "observable." Moreover, the line of distinction between the "theoretical" and the "observable" is always diffuse and to some extent arbitrary (see Feyerabend, "An Attempt at a Realistic Interpretation of Experience" and Maxwell, Meaning Postulates in Scientific Theories).

One critical problem remains. We have seen that, as Carnap has emphasized [ESO], our choice of a framework will, in general, be influenced by theoretical (factual) knowledge. The question arises: How is such knowledge to be expressed? The answer is, obviously, in some *framework* or other, and it must be one which is already in use. Thus there can never be any such thing as the abrupt, wholesale adoption of a completely new framework, nor can there ever be the abrupt wholesale abandonment of a framework presently being employed. A "new" framework will always consist of at least a portion of an "old" one *plus* new rules, new A-truths, and so forth. The modification of a framework (and, actually, we should only speak of *modification* rather than of *adoption* of "new" frameworks) is a "bootstraps" operation. A portion of it will be used to give arguments,[12] some of which will usually contain factual premises, as to why another portion of it should be enriched, modified, or abandoned. Even when we are assessing a framework or portion thereof for the *condition of adequacy,* we must in so doing employ *some* framework or portion thereof, the choice of which, in its turn can have been made only on the basis of *its* adequacy, and, of necessity will have been influenced by some "purely pragmatic" factors (preferences arising from past linguistic experience, if nothing else). This shows both why the epistemological enterprise must be to some extent circular and why the *condition of adequacy* can never be completely extricated from *purely pragmatic* considerations. I am sorry

[11] See, e.g., H. Feigl, "Existential Hypotheses," *Philosophy of Science* 17:35–62 (1950); Wilfrid Sellars, "The Language of Theories" in CIPS; Paul K. Feyerabend, "An Attempt at a Realistic Interpretation of Experience," *Proc. Arist. Soc.* 144–70 (1958); M. Scriven, "Definitions, Explanations, and Theories," in MSPS, Vol. II; and my "The Ontological Status of Theoretical Entities," MSPS, Vol. III.

[12] Since the language we first learn to speak is (I suppose, by *definition*) "ordinary language," it follows that when we *first* begin to modify and enrich our framework, whether by scientific theorizing or by philosophical analysis, both the *examined* portion and the portion of the framework with which we *do the examining* will be parts of the "ordinary" framework. And for many purposes, a slight modification of this "ordinary" framework often suffices. But it should also be obvious that by continuing the process of piecemeal modification, whereby we add new terms and new A-truths and abandon old ones, it would be possible, in principle, to completely replace the "ordinary" (or any other) framework. (Cf. G. Maxwell and H. Feigl, "Why Ordinary Language Needs Reforming," *Journal of Philosophy,* V. 58, no. 18 (August 31, 1961).

that we can never quite reach rock bottom in the vindication of a framework and that we cannot stand back in metaphysical grandeur and make our philosophical decisions independently of all frameworks, but, as Bertrand Russell says, it is not my fault.

As far as theoretical entities are concerned, the dénouement should be obvious. The A-true sentences mentioning such entities tell how they *must* (within a given framework) and the contingent postulates mentioning them tell us how they *may* (if the postulates are true, how they *as a matter of fact do*) *differ from* and *resemble* entities with which we are perhaps more familiar. No metaphysical questions concerning their nature or their "reality status" are left over. If there are highly confirmed sentences which entail the sentence 'There are ϕs', where ϕs are a kind of theoretical entities, then it is highly confirmed that ϕs enjoy just as "full-blown existence"—just as "much reality"—as any kind of entities whatever.

Consider—a rather extreme example—lines of force in, say, a magnetic field. If we have theories which are highly confirmed and which, together with other true (or highly confirmed) contingent statements, entail that there are lines of force, then it is highly confirmed that lines of force exist (are real). True, they are not physical objects (neither, I suppose, are photons or, for that matter, shadows or pains in the big toe); the very A-truths involving them tell us this; and the total set of such A-truths should completely dispel any ontological anxieties we may feel concerning them. It is true that, our theories being what they are, we may not be quite sure as to what the most felicitously selected set of such A-truths would be. But this in no way suggests that lines of force are "convenient fictions"—any more than are isobars, the *actual* lines of equal elevation corresponding to contour lines on a map, temperature gradients, and so forth.

If someone would impugn the status of theoretical entities, it is incumbent upon him to produce a framework which fulfills the *condition of adequacy* and which does not mention such entities or, at the very least, to show that (suitably reconstructed) theories which are extant today do *not* fulfill this condition; he should also explain how it is that some entities which were once "theoretical" are today "observable," for example microbes, certain viruses (via the electron microscope), and so forth.[18]

University of Minnesota

---

[18] It might seem, superficially, that the central theses of this paper hinge on a sharp distinction between A-truth and contingency. However, I believe that they can easily be accommodated to those who prefer a reconstruction of, say, the Quinian type, on the one hand, or a (Wilfrid) Sellarsian kind, on the other. In such reconstructions, any law-like statement would play a double role. It would be a "framework principle" and, thus, part of the implicit definition of the non-logical terms occurring in it, and it would also express a factually non-empty law.

HILARY PUTNAM

What Theories Are Not

~~~~~~

The announced topic for this symposium was the role of models in empirical science; however, in preparing for this symposium, I soon discovered that I had first to deal with a different topic, and this different topic is the one to which this paper actually will be devoted. The topic I mean is the role of *theories* in empirical science; and what I do in this paper is attack what may be called the 'received view' on the role of theories—that theories are to be thought of as 'partially interpreted calculi' in which only the 'observation terms' are 'directly interpreted' (the theoretical terms being only 'partially interpreted', or, some people even say, 'partially understood').

To begin, let us review this received view. The view divides the non-logical vocabulary of science into two parts:

OBSERVATION TERMS	THEORETICAL TERMS
such terms as	such terms as
'red',	'electron',
'touches',	'dream',
'stick', etc.	'gene', etc.

The basis for the division appears to be as follows: the observation terms apply to what may be called publicly observable things and signify observable qualities of these things, while the theoretical terms correspond to the remaining unobservable qualities and things.

Reprinted from *Logic, Methodology and Philosophy of Science,* ed. by Ernest Nagel, Patrick Suppes, and Alfred Tarski with the permission of the author and the publishers, Stanford University Press. Copyright 1962 by the Board of Trustees of the Leland Stanford Junior University.

This division of terms into two classes is then allowed to generate a division of statements into two[1] classes as follows:

OBSERVATIONAL STATEMENTS	THEORETICAL STATEMENTS
statements containing only observation terms and logical vocabulary	statements containing theoretical terms

Lastly, a scientific theory is conceived of as an axiomatic system which may be thought of as initially uninterpreted, and which gains 'empirical meaning' as a result of a specification of meaning *for the observation terms alone*. A kind of partial meaning is then thought of as drawn up to the theoretical terms, by osmosis, as it were.

THE OBSERVATIONAL-THEORETICAL DICHOTOMY

One can think of many distinctions that are crying out to be made ('new' terms vs. 'old' terms, technical terms vs. non-technical ones, terms more-or-less peculiar to one science vs. terms common to many, for a start). My contention here is simply:

(1) The *problem* for which this dichotomy was invented ('how is it possible to interpret theoretical terms?') does not exist.

(2) A basic reason some people have given for introducing the dichotomy is false: namely, justification in science does *not* proceed 'down' in the direction of observation terms. In fact, justification in science proceeds in any direction that may be handy—more observational assertions sometimes being justified with the aid of more theoretical ones, and vice versa. Moreover, as we shall see, while the notion of an *observation report* has some importance in the philosophy of science, such reports cannot be identified on the basis of the vocabulary they do or do not contain.

(3) In any case, whether the reasons for introducing the dichotomy were good ones or bad ones, the double distinction (observation terms —theoretical terms, observation statements—theoretical statements) presented above is, in fact, completely broken-backed. This I shall try to establish now.

In the first place, it should be noted that the dichotomy under discussion was intended as an explicative and not merely a stipulative one. That is, the words 'observational' and 'theoretical' are not having arbitrary new meanings bestowed upon them; rather, pre-existing uses of these words (especially in the philosophy of science) are presumably

---

[1] Sometimes a *tripartite* division is used: observation statements, theoretical statements (containing *only* theoretical terms), and 'mixed' statements (containing both kinds of terms). This refinement is not considered here, because it avoids none of the objections presented below.

being sharpened and made clear. And, in the second place, it should be recalled that we are dealing with a double, not just a single, distinction. That is to say, part of the contention I am criticizing is that, once the distinction between observational and theoretical *terms* has been drawn as above, the distinction between theoretical statements and observational reports or assertions (in something like the sense usual in methodological discussions) can be drawn in terms of it. What I mean when I say that the dichotomy is 'completely broken-backed' is this:

(A) If an 'observation term' is one that cannot apply to an unobservable, then there are no observation terms.[2]

(B) Many terms that refer primarily to what Carnap would class as 'unobservables' are not theoretical terms; and at least some theoretical terms refer primarily to observables.

(C) Observational reports can and frequently do contain theoretical terms.

(D) A scientific theory, properly so-called, may refer only to observables. (Darwin's theory of evolution, as originally put forward, is one example.)

To start with the notion of an 'observation term': Carnap's formulation in *Testability and Meaning* [1] was that for a term to be an observation term not only must it correspond to an observable quality, but the determination whether the quality is present or not must be able to be made by the observer in a relatively short time, and with a high degree of confirmation. In his most recent authoritative publication [2], Carnap is rather brief. He writes, 'the terms of V_0 [the 'observation vocabulary'—H.P.] are predicates designating observable properties of events or things (e.g., 'blue', 'hot', 'large', etc.) or observable relations between them (e.g., '*x* is warmer than *y*, '*x* is contiguous to *y*', etc.)' [2, p. 41)]. The only other clarifying remarks I could find are the following: 'The name 'observation language' may be understood in a narrower or in a wider sense; the observation language in the wider sense includes the disposition terms. In this article I take the observation language L_0 in the narrower sense' [2, p. 63]. 'An observable property may be regarded as a simple special case of a testable disposition: for example, the operation for finding out whether a thing is blue or hissing or cold, consists simply in looking or listening or touching the thing, respectively. Nevertheless, *in the reconstruction of the language* [italics mine—H.P.] it seems convenient to take some properties for which the test procedure is extremely simple (as in the

---

[2] I neglect the possibility of trivially constructing terms that refer only to observables: namely, by conjoining 'and is an observable thing' to terms that would otherwise apply to some unobservables. 'Being an observable thing' is, in a sense, highly theoretical and yet applies only to observables!

examples given) as directly observable, and use them as primitives in Lo'
[2, p. 63].

These paragraphs reveal that Carnap, at least, thinks of observation
terms as corresponding to qualities that can be detected without the
aid of instruments. But always so detected? Or can an observation term
refer sometimes to an observable thing and sometimes to an unobservable? While I have not been able to find any explicit statement on this
point, it seems to me that writers like Carnap must be *neglecting* the
fact that *all* terms—including the 'observation terms'—have at least the
possibility of applying to unobservables. Thus their problem has sometimes been formulated in quasi-historical terms—'How could theoretical terms have been introduced into the language?' And the usual discussion strongly suggests that the following puzzle is meant: if we
imagine a time at which people could only talk about observables (had
not available any theoretical terms), how did they ever manage to *start*
talking about unobservables?

It is possible that I am here doing Carnap and his followers an in-
justice. However, polemics aside, the following points must be emphasized:

(1) Terms referring to unobservables are *invariably* explained, in
the actual history of science, with the aid of already present locutions
referring to unobservables. There never was a stage of language at
which it was impossible to talk about unobservables. Even a three-year-
old child can understand a story about 'people too little to see' [3] and
not a single 'theoretical term' occurs in this phrase.

(2) There is not even a single *term* of which it is true to say that it
could not (without changing or extending its meaning) be used to refer
to unobservables. 'Red', for example, was so used by Newton when he
postulated that red light consists of *red corpuscles*.[4]

In short: if an 'observation term' is a term which *can*, in principle,
only be used to refer to observable things, then *there are no observa-*

---

[3] Von Wright has suggested (in conversation) that this is an *extended* use of language (because we first learn words like "people" in connection with people we *can*
see). This argument from "the way we learn to use the word" appears to be un-
sound, however (cf. [4]).

[4] Some authors (although not Carnap) explain the intelligibility of such discourse
in terms of logically possible submicroscopic observers. But (a) such observers could
not see single photons (or light corpuscles) even on Newton's theory; and (b) once
such physically impossible (though logically possible) 'observers' are introduced, why
not go further and have observers with sense organs for electric charge, or the curvature of space, etc.! Presumably because *we* can see *red*, but not *charge*. But then,
this just makes the point that we *understand* 'red' even when applied outside our
normal 'range', even though we learn it ostensively, without *explaining* that fact.
(The explanation lies in this: that understanding any term—even "red"—involves
at least two elements: internalizing the syntax of a natural language, and acquiring
a background of ideas. Overemphasis on the way 'red' is *taught* has led some
philosophers to misunderstand how it is *learned*.)

tion terms. If, on the other hand, it is granted that locutions consisting of just observation terms can refer to unobservables, there is no longer any reason to maintain *either* that theories and speculations about the unobservable parts of the world must contain 'theoretical (= non-observation) terms' *or* that there is any general problem as to how one can introduce terms referring to unobservables. Those philosophers who find a difficulty in how we understand theoretical terms should find an equal difficulty in how we understand 'red' and 'smaller than'.

So much for the notion of an 'observation term'. Of course, one may recognize the point just made—that the 'observation terms' also apply, in some contexts, to unobservables—and retain the class (with a suitable warning as to how the label 'observation term' is to be understood). But can we agree that the complementary class—what should be called the 'non-observation terms'—is to be labelled 'theoretical terms'? No, for the identification of 'theoretical term' with 'term (other than the 'disposition terms', which are given a special place in Carnap's scheme) designating an unobservable quality' is unnatural and misleading. On the one hand, it is clearly an enormous (and, I believe, insufficiently motivated) extension of common usage to classify such terms as 'angry', 'loves', and so forth, as 'theoretical terms' simply because they allegedly do not refer to public observables. A theoretical term, properly so-called, is one which comes from a scientific *theory* (and the almost untouched problem, in thirty years of writing about 'theoretical terms' is what is *really* distinctive about such terms). In this sense (and I think it the sense important for discussions of science) 'satellite' is, for example, a theoretical term (although the things it refers to are quite observable[5]) and 'dislikes' clearly is not.

Our criticisms so far might be met by re-labelling the first dichotomy (the dichotomy of terms) 'observation vs. non-observation', and suitably 'hedging' the notion of 'observation'. But more serious difficulties are connected with the identification upon which the second dichotomy is based—the identification of 'theoretical statements' with statements containing non-observation ('theoretical') terms, and 'observation statements' with 'statements in the observational vocabulary'.

That observation statements may contain theoretical terms is easy to establish. For example, it is easy to imagine a situation in which the following sentence might occur: 'We also *observed* the creation of two electron-positron pairs'.

This objection is sometimes dealt with by proposing to 'relativize'

---

[5] Carnap might exclude 'satellite' as an observation term, on the ground that it takes a comparatively long time to verify that something is a satellite with the naked eye, even if the satellite is close to the parent body (although this could be debated). However, 'satellite' cannot be excluded on the quite different ground that many satellites are too far away to see (which is the ground that first comes to mind) since the same is true of the huge majority of all *red* things.

the observation-theoretical dichotomy to the context. (Carnap, however, rejects this way out in the article we have been citing.) This proposal to 'relativize' the dichotomy does not seem to me to be very helpful. In the first place, one can easily imagine a context in which 'electron' would occur, in the same text, in *both* observational reports and in theoretical conclusions from those reports. (So that one would have distortions if one tried to put the term in either the 'observational term' box or in the 'theoretical term' box.) In the second place, for what philosophical problem or point does one require even the relativized dichotomy?

The usual answer is that sometimes a statement A (observational) is offered in support of a statement B (theoretical). Then, in order to explain why A is not itself questioned in the context, we need to be able to say that A is functioning, in that context, as an observation report. But this misses the point I have been making! I do not deny the need for some such notion as 'observation report'. What I deny is that the distinction between observation reports and, among other things, theoretical statements, can or should be drawn on the basis of vocabulary. In addition, a relativized dichotomy will not serve Carnap's purposes. One can hardly maintain that theoretical terms are only partially interpreted, whereas observation terms are completely interpreted, if no sharp line exists between the two classes. (Recall that Carnap takes his problem to be 'reconstruction of the language', not of some isolated scientific context.)

PARTIAL INTERPRETATION

The notion of 'partial interpretation' has a rather strange history—the term certainly has a technical ring to it, and someone encountering it in Carnap's writings, or Hempel's, or mine[6] certainly would be justified in supposing that it was a term from mathematical logic whose exact definition was supposed to be too well known to need repetition. The sad fact is that this is not so! In fact, the term was introduced by Carnap in a section of his monograph [3], without definition (Carnap *asserted* that to interpret the observation terms of a calculus is automatically to 'partially interpret' the theoretical primitives, without explanation), and has been subsequently used by Carnap and other

[6] I used this notion uncritically in [5]. From the discussion, I seem to have had concept (2) (below) of 'partial interpretation' in mind, or a related concept. (I no longer think *either* that set theory is helpfully thought of as a 'partially interpreted calculus' in which only the 'nominalistic language' is directly interpreted, *or* that mathematics is best identified with set theory for the purposes of philosophical discussion, although the idea that certain statements of set theory, e.g., the continuum hypothesis, do not have a defined truth-value does have a certain appeal, given the unclarity of our notion of a 'set'.)

authors including myself) with copious cross references, but with no further explanation.

One can think of (at least) three things that 'partial interpretation' could mean. I will attempt to show that none of these three meanings is of any use in connection with the 'interpretation of scientific theories'. My discussion has been influenced by a remark of Ruth Anna Mathers to the effect that not only has the concept been used without any *definition*, but it has been applied indiscriminately to *terms, theories*, and *languages*.

(1) One might give the term a meaning from mathematical logic as follows (I assume familiarity here with the notion of a 'model' of a formalized theory): to 'partially interpret' a theory is to specify a nonempty class of intended models. If the specified class has one member, the interpretation is *complete*; if more than one, properly *partial*.

(2) To partially interpret a *term P* could mean (for a Verificationist, as Carnap is) to specify a verification-refutation procedure. If $\bar{a}_1$ is an individual constant designating an individual a_1 (Carnap frequently takes space-time points as the individuals, assuming a 'field' language for physics), and it is possible to verify $P(\bar{a}_1)$, then the individual a_1 is in the extension of the term P; if $P(\bar{a}_1)$ is refutable, then a_1 is in the extension of $\bar{P}$, the negation of P; and if the existing test procedures do not apply to a_1 (e.g., if a_1 fails to satisfy the antecedent conditions specified in the test procedures) then it is *undefined* if a_1 is in or out of the extension of P.

This notion of partial interpretation of *terms* immediately applies to terms introduced by reduction sentences[7] (Carnap calls these 'pure disposition terms'). In this case the individual a_1 is either in the extension of P or in the extension of $\bar{P}$, provided the antecedent of at least one reduction sentence 'introducing' the term P is true of a_1, and otherwise it is *undefined* whether $P(a_1)$ is true or not. But it can be extended to theoretical primitives in a *theory* as follows: If $P(\bar{a}_1)$ follows from the postulates and definitions of the theory and/or the set of all true observation sentences, then a_1 is in the extension of P; if $\bar{P}(\bar{a}_1)$ follows from the postulates and definitions of the theory and/or the set of all true observation sentences, then a_1 is in the extension of $\bar{P}$; in all other cases, $P(\bar{a}_1)$ has an *undefined* truth-value.

(3) Most simply, one might say that to partially interpret a formal *language* is to *interpret part* of the language (e.g., to provide translations into common language for some terms and leave the others mere dummy symbols).

Of these three notions, the first will not serve Carnap's purposes, because it is necessary to use theoretical terms in order to specify even

---

[7] For the definition of this concept, see [1].

a *class* of intended models for the usual scientific theories. Thus, consider the problem of specifying the intended values of the individual variables. If the language is a 'particle' language, then the individual variables range over 'things'—but things in a *theoretical* sense, including mass points and systems of mass points. It is surely odd to take the notion of a 'physical object' as either an observational or a purely logical one when it becomes wide enough to include point-electrons, at one extreme, and galaxies at the other. On the other hand, if the language is a 'field' language, then it is necessary to say that the individual variables range over *space-time points*—and the difficulty is the same as with the notion of a 'physical object.'

Moving to the predicate and function-symbol vocabulary: consider, for example, the problem of specifying either a unique intended interpretation or a suitable class of models for Maxwell's equations. We must say, at least, that E and H are intended to have as values vector-valued functions of *space-time points,* and that the norms of these vectors are to measure, roughly speaking, the velocity-independent force on a small test particle per unit charge, and the velocity-dependent force per unit charge. One might identify force with mass times (a suitable component of) acceleration, and handle the reference to an (idealized) test particle via 'reduction sentences', but we are still left with 'mass', 'charge', and, of course, 'space-time point'. ('Charge' and 'mass' have as values a real-valued function and a non-negative real-valued function, respectively, of space-time points; and the values of these functions are supposed to measure the intensities with which certain *physical magnitudes* are present at these points—where the last clause is necessary to rule out flagrantly unintended interpretations that can never be eliminated otherwise.)

(One qualification: I said that *theoretical* terms are necessary to specify even a *class* of intended models—or of models that a realistically minded scientist could accept as what he has in mind. But 'physical object', 'physical magnitude' and 'space-time point' are not—except for the last—'theoretical terms', in any idiomatic sense, any more than they are 'observation terms'. Let us call them for the nonce simply 'broad spectrum terms'—noting that they pose much the same problems as do certain meta-scientific terms, for example, 'science' itself. Of them we might say, as Quine does of the latter term [6], that they are not defined in advance—rather science itself tells us (with many changes of opinion) what the scope of 'science' is, or of an individual science, for example, chemistry, what an 'object' is, what 'physical magnitudes' are. In this way, these terms, although not theoretical terms, tend eventually to acquire technical senses via theoretical definitions.)

A further difficulty with the first notion of 'partial interpretation' is that theories with false observational consequences have *no* interpretation (since they have no model that is 'standard' with respect to the observation terms). This certainly flies in the face of our usual notion of an interpretation, according to which such a theory is *wrong*, not *senseless*.

The second notion of partial interpretation that we listed appears to me to be totally inadequate even for the so-called 'pure disposition terms', for example, 'soluble'. Thus, let us suppose, for the sake of a simplified example, that there were only one known test for *solubility* —immersing the object in water. Can we really accept the conclusion that it has a *totally undefined* truth-value to say of something that is never immersed in water that it is soluble?

Let us suppose now that we notice that all the sugar cubes that we immerse in water dissolve. On the basis of this *evidence* we *conclude* that all sugar is soluble—even cubes that are never immersed. On the view we are criticizing, this has to be described as 'linguistic stipulation' rather than as 'discovery'! Namely, according to this concept of partial interpretation, what we do is *give* the term 'soluble' the *new* meaning 'soluble-in-the-old-sense-of-sugar'; and what we ordinarily describe as evidence that the un-immersed sugar cubes are soluble should rather be described as evidence that our new meaning of the term 'soluble' is compatible with the original 'bilateral reduction sentence'.

Also, although it will now be true to say 'sugar is soluble', it will still have a totally undefined truth-value to say of many, say, lumps of *salt* that *they* are soluble.

Ordinarily, 'change of meaning' refers to the sort of thing that happened to the word 'knave' (which once meant 'boy'), and 'extension of meaning' to the sort of thing that happened in Portugal to the word for family ('familhia'), which has come to include the servants. In these senses, which also seem to be the only ones useful for linguistic theory, it is simply *false* to say that in the case described (concluding that sugar is soluble) the word 'soluble' underwent either a change or an extension of meaning. The *method of verification* may have been extended by the discovery, but this is only evidence that method of verification is not meaning.

In any case, there does not seem to be any reason why we cannot agree with the customary account. What we meant all along by 'it's soluble' was, of course, 'if it *were* in water, it would dissolve'; and the case we described can *properly* be described as one of drawing an inductive inference—to the conclusion that all these objects (lumps of sugar, whether immersed or not) are soluble in *this* sense. Also, there is no reason to reject the view, which is certainly built into our use of the term 'soluble', that it has a definite (but sometimes unknown) truth-value to say of anything (of a suitable size and consistency) that it is soluble, whether it satisfies a presently-known test condition or not. Usually the objection is raised that it is 'not clear what it means' to say 'if it *were* in water, it would dissolve'; but there is no *linguistic* evidence of this unclarity. (Do people construe it in different ways? Do they ask for a paraphrase? Of course, there is a philosophical problem having to do with 'necessary connection', but one must not confuse a word's being connected with a philosophical problem with its possessing an unclear meaning.)

Coming now to theoretical terms (for the sake of simplicity, I assume that our world is non-quantum mechanical): If we want to preserve the ordinary world-picture at all, we certainly want to say it has a definite truth-value to say that there is a helium atom inside any not-too-tiny region X. But in fact, our test conditions—even if we allow tests implied by a theory, as outlined above under (2)—will not apply to small regions X in the interior of the sun, for example (or in the interior of many bodies at many times). Thus we get the following anomalous result: it is *true* to say that there are helium atoms in the sun, but it is neither true nor false that one of these is inside any given tiny subregion X! Similar things will happen in connection with theoretical statements about the very large, for example, it may be 'neither true nor false' that the average curvature of space is positive, or that the universe is finite. And once again, perfectly ordinary scientific discoveries will constantly have to be described as 'linguistic stipulations', 'extensions of meaning', and so forth.

Finally, the third sense of 'partial interpretation' leads to the view that theoretical terms have *no meaning at all,* that they are mere computing devices, and is thus unacceptable.

To sum up: We have seen that of the three notions of 'partial interpretation' discussed, each is either unsuitable for Carnap's purposes (starting with observation terms), or incompatible with a rather minimal scientific realism; and, in addition, the second notion depends upon gross and misleading changes in our use of language. Thus in *none* of these senses is 'a partially interpreted calculus in which only the observation terms are directly interpreted' an acceptable model for a scientific theory.

INTRODUCING THEORETICAL TERMS

We have been discussing a proposed solution to a philosophical problem. But what *is* the problem?

The problem is sometimes referred to as the problem of 'interpreting', that is, giving the meaning of theoretical terms in science. But this cannot be much of a *general* problem (it may, of course, be a problem in specific cases). Why should not one be able to give the meaning of a theoretical term? (Using, if necessary, *other* theoretical terms, 'broad spectrum' terms, etc.) The problem might be restated— to give the meaning of theoretical terms, *using only observation terms.*

But then, why should we suppose that this is or ought to be possible? Something like this may be said: suppose we make a 'dictionary' of theoretical terms. If we allow theoretical terms to appear both as 'entries' and in the *definitions,* then there will be 'circles' in our dictionary. But there are circles in every dictionary!

We perhaps come closer to the problem if we observe that, while dictionaries are useful, they are useful only to speakers who already know a good deal of the language. One cannot learn one's native language to begin with from a dictionary. This suggests that the problem is really to give an account of how the use of theoretical terms is *learned* (in the life-history of an individual speaker); or, perhaps, of how theoretical terms are 'introduced' (in the history of the language).

To take the first form of the problem (the language-learning of the individual speaker): It appears that theoretical terms are learned in essentially the way most words are learned. Sometimes we are given lexical definitions (e.g., 'a *tiglon* is a cross between a tiger and a lion'); more often, we simply imitate other speakers; many times we combine these (e.g., we are given a lexical definition, from which we obtain a rough idea of the use, and then we bring our linguistic behavior more closely into line with that of other speakers via imitation).

The story in connection with the introduction of a new technical term into the *language* is roughly similar. Usually, the scientist introduces the term via some kind of paraphrase. For example, one might explain 'mass' as 'that physical magnitude which determines how strongly a body resists being accelerated, e.g., if a body has twice the mass it will be twice as hard to accelerate'. (Instead of 'physical magnitude' one might say, in ordinary language, 'that property of the body', or 'that *in* the body which. . . . Such 'broad-spectrum' notions occur in every natural language; and our present notion of a 'physical magnitude' is already an extreme refinement.) Frequently, as in the case of 'force' and 'mass', the term will be a common-language term whose new technical use is in some respects quite continuous with the ordinary use. In such cases, a lexical definition is frequently omitted, and in its place one has merely a statement of some of the differences between the usual use and the technical use being introduced. Usually one gains only a rough idea of the use of a technical term from these explicit metalinguistic statements, and this rough idea is then refined by reading the theory or text in which the term is employed. However, the role of the explicit metalinguistic statement should not be overlooked: one could hardly read the text or technical paper with understanding if one had neither explicit metalinguistic statements or previous and related uses of the technical words to go by.

It is instructive to compare here the situation with respect to the logical connectives, in their modern technical employment. We introduce the precise and technical senses of 'or', 'not', 'if-then', and so forth, using the imprecise 'ordinary language' *or, and, not* and so forth. For example, we say '$A \lor B$ shall be true *if A* is true, *and* true *if B* is true, *and* $A \lor B$ shall be false *if A* is false *and B* is false. In particular, $A \lor B$ shall be true *even if A and B* are *both* true.' Notice that no one has proposed to say that '$\lor$' is only 'partially interpreted' because we use 'and', 'if', etc., in the ordinary imprecise way when we 'introduce' it.

In short, we can and do perform the feat of using imprecise language to introduce more precise language. This is like all use of tools—we use less-refined tools to manufacture more-refined ones. Secondly, there are even ideas that can be expressed in the more precise language that could not be intelligibly expressed in the original language. Thus, to borrow an example due to Alonzo Church, a statement of the form $(((A \supset B) \supset B) \supset B)$ can probably not be intelligibly rendered in ordinary language—although one can understand it once one has had an explanation of '$\supset$' in ordinary language.

It may be, however, that the problem is supposed to be *this:* to *formalize* the process of introducing technical terms. Let us try this problem out on our last example (the logical connectives). Clearly we *could* formalize the process of introducing the usual truth-functional connectives. We would only have to take as primitives the 'ordinary language' *and, or, not* in their usual (imprecise) meanings, and then we could straightforwardly write down such characterizations as the one given above for the connective '$\lor$'. But if someone said: 'I want you to introduce the logical connectives, quantifiers, and so forth, without having any *imprecise* primitives (because using imprecise notions is not 'rational reconstruction') and also without having any *precise* logical symbols as primitives (because that would be 'circular')', we should just have to say that the task was impossible.

The case appears to me to be much the same with the 'theoretical terms'. If we take as primitives not only the 'observation terms' and the 'logical terms', but also the 'broad-spectrum' terms referred to before ('thing', 'physical magnitude', etc.), and, perhaps, certain imprecise but useful notions from common language—for example, '*harder to accelerate*', 'determines'—then we can introduce theoretical terms without difficulty:

(1) Some theoretical terms can actually be explicitly defined in Carnap's 'observation language'. Thus, suppose we had a theory according to which everything consisted of 'classical' elementary particles —little-extended individual particles; and suppose no two of these were supposed to touch. Then 'elementary particle', which is a 'theoretical term' if anything is, would be explicitly definable: X is an elementary particle $\equiv X$ cannot be decomposed into parts Y and Z which are not contiguous—and the above definition requires only the notions 'X is a part of Y' and 'X is contiguous to Y'. (If we take *contiguity* as

a reflexive relation, then 'part of' is definable in terms of it: X is a part of $Y \equiv$ everything that is contiguous to X is contiguous to Y. Also, Y and Z constitute a 'decomposition' of X if (i) nothing is a part of both Y and Z; (ii) X has no part which contains no part in common with either Y or Z. However, it would be perfectly reasonable, in my opinion, to take 'part of' as a *logical* primitive, along with 'is a member of'—although Carnap would probably disagree.)

We note that the, at first blush surprising, possibility of defining the obviously theoretical term 'elementary particle' in Carnap's 'observation language' rests on the fact that the notion of a *physical object* is smuggled into the language in the very interpretation of the individual variables.

(2) The kind of characterization we gave above for 'mass' (using the notion 'harder to accelerate') could be formalized. Again a broad-spectrum notion ('physical magnitude') plays a role in the definition.

But once again, no one would normally want to formalize such obviously informal definitions of theoretical terms. And once again, if someone says: 'I want you to introduce the theoretical terms using *only* Carnap's *observation terms*', we have to say, apart from special cases (like that of the 'classical' notion of an elementary particle), that this seems impossible. But why should it be possible? And what philosophic moral should we draw from the impossibility?—Perhaps only this: that we are able to have as rich a theoretical vocabulary as we do have because, thank goodness, we were never in the position of having *only* Carnap's observation vocabulary at our disposal.

REFERENCES

CARNAP, R. Testability and meaning. Pp. 47–92 in *Readings in the Philosophy of Science,* H. Feigl and M. Brodbeck, eds., New York, Appleton-Century-Crofts, 1955, x+517 pp. Reprinted from *Philosophy of Science,* Vol. 3 (1936) and Vol. 4 (1937).

CARNAP, R. The methodological character of theoretical concepts. Pp. 1–74 in *Minnesota Studies in the Philosophy of Science,* I., H. Feigl *et al.,* eds., Minneapolis, University of Minnesota Press, 1956, x+517 pp.

CARNAP, R. *The Foundations of Logic and Mathematics.* Vol. 4, no. 3 of the International Encyclopedia of Unified Science, Chicago, University of Chicago Press, 1939, 75 pp.

FODOR, J. Do words have uses? Submitted to *Inquiry.*

PUTNAM, H. Mathematics and the existence of abstract entities. *Philosophical Studies,* Vol. 7 (1957), 81–88.

QUINE, W. V. O. The scope and language of science, *British Journal for the Philosophy of Science,* Vol. 8 (1957), 1–17.

Review of Putnam

2. MODELS

The term 'model' is often used where 'theory' would do as well; Braithwaite holds that in a distinctive and interesting sense,*

> . . . *a model for a theory T* is another theory *M* which corresponds to the theory *T* in respect of deductive structure . . . an alternative and equivalent explication of *model for a theory* can be given by saying that a model is another interpretation of the theory's calculus (225).

The second formulation seems to be in keeping with Suppes' statement, that "the logical notion of a model of a theory . . . is the fundamental one for the empirical sciences as well as mathematics" (252). However, when Braithwaite characterizes the *contextualist* account of the functioning of theoretical concepts, he speaks of "an interpretation of the calculus expressing the theory which works from the bottom upwards" (230).

> The final theorems of the calculus are interpreted as expressing empirically testable generalizations, the axioms of the calculus are interpreted as propositions from which these generalizations logically follow, and the theoretical terms occurring in the calculus are given a meaning implicitly by their context, i.e., by their place within the calculus (230).

Reprinted and excerpted from the Journal of Philosophy, Vol. LXI, No. 2 (January 16, 1964), 80–84, by kind permission of the author and publishers.
* The reference is to R. B. Braithwaite, "Models in the Empirical Sciences," in *Logic, Methodology and Philosophy of Science,* edited by E. Nagel, P. Suppes, and A. Tarski, Stanford, Calif.: Stanford University Press, 1962; but the view under discussion is essentially similar to that in the Braithwaite article reprinted in this anthology, pp. 47–52.

Putnam is at pains to point out that such a notion of *partial interpretation* of a calculus, like the distinction between observational and theoretical terms, is a dim one. Putnam's attack is directed at

> . . . the "received view" on the role of theories—that theories are to be thought of as "partially interpreted calculi" in which only the "observation terms" are "directly interpreted" (the theoretical terms being only "partially interpreted," or, some people even say, "partially understood") (240).

Putnam's claims concerning the theoretical-observational dichotomy are these:

> (1) The *problem* for which this dichotomy was invented ("how is it possible to interpret theoretical terms?") does not exist.
>
> (2) A basic reason some people have given for introducing the dichotomy is false: namely, justification in science does *not* proceed "down" in the direction of observation terms. . . . Moreover, . . . while the notion of an *observation report* has some importance in the philosophy of science, such reports cannot be identified on the basis of the vocabulary they do or do not contain.
>
> (3) In any case, . . . the double distinction (observation terms-theoretical terms, observation statements-theoretical statements) . . . is, in fact, completely broken-backed because. . . .
>
> (A) If an "observation term" is one that cannot apply to an unobservable then there are no observation terms.
>
> (B) Many terms that refer primarily to what Carnap would class as "unobservables" are not theoretical terms; and at least some theoretical terms refer primarily to observables.
>
> (C) Observational reports can and frequently do contain theoretical terms.
>
> (D) A scientific theory, properly so-called, may refer only to observables (241).

I should like to comment on some important issues which are raised here.

3. OBSERVATION REPORTS

As Putnam says, these cannot be classified on the basis of the vocabulary they do or do not contain. To this it should be added that sentences as such cannot be classified as observation reports or not. The point is not that sentences can be used by different people (or at different times, or in different contexts) to make different *statements,* which are accordingly the appropriate subjects of the predicate, *is an observation report.* Rather, the claim is that to make sense of the one-place predicate, *is an observation report,* it must be applied to *utterances,* that is, to sentence tokens, construed as events: to sentences in the mouths of speakers. Alternatively, the work of such a one-place predicate could be done by a ternary relation,

(sentence) s is an observation report by (person) X at (time) t.

Thus, $s =$ 'Moriarty is hungry' might be an observation report by X at a time when it is also uttered as an inferred truth by X', who can not see and hear Moriarty attack his food, but who knows that Moriarty is beginning his first meal since breakfast, twelve hours ago, and is in good health.

The notion of a sentence's *being observational* seems to be derivative from the notion of its *being an observation report* for a person at a time. A sentence is *observational* if, in some rather general sense of 'possible', it is possible for there to be a person and a time relative to which s, uttered by the person at the time, should be an observation report. On the other hand, it seems that not even such a loose characterization of *observation report* can be gotten from *observational sentence*. An observation report (by X at t) is a sentence, uttered by X at t, which X knows to be true on the basis of what he has observed at t or shortly before. It adds nothing to this characterization of observation reports, to stipulate that the sentence which X utters must be an observational sentence.

From these remarks it does not follow that the *observational sentences* cannot be identified on the basis of the vocabulary they do or do not contain; if they can, there may be a useful basis for the distinction between *terms* that are observational and those which are not, and it may be appropriate to call the latter "theoretical."

4. OBSERVABLES

We do classify physical objects as *observed* or not (by a person at a time), and even as *observable* or not; but these classifications need not have much importance for the philosophy of science. The notions, *observed* and *observable,* can be applied to events with much the same effect that is obtained by applying *is an observation report* and *is observational* to sentences. But, like Putnam in (A), (B), and (D), I am concerned here with the observability, not of events, but of such particulars as roses, people, shadows, numbers, and electrons. The opposition here is between *observables* and *theoretical constructs,* and Putnam shows that, however this opposition may be understood, it cannot be explained as the difference between the individuals to which observational predicates meaningfully apply and those to which theoretical predicates meaningfully apply; nor can the observables be identified as those individuals which are referred to by terms that have essential occurrences in observation reports; nor need a theory refer to nonobservables.

In (A), (B), and (D), Putnam uses the notion of an *observable individual* to undermine the notion of an *observational predicate*. His arguments show that one must give up one of these notions, or at least give up the claim that they are related in the ways that are attacked in (A), (B), and (D). In these three statements Putnam seems to accept the notion of an *observable individual* as a clear and useful one: useful at least as a surgical tool, to excise the corrupt concept of an *observational predicate* from the philosophy of science. But the arguments can easily be turned around, to show that the notion of an *observable individual* is in the same case. For (C) might well be altered to read:

> (C′) Observational reports can generally be interpreted so that all singular terms refer to entities which would ordinarily be called "theoretical" or "nonobservational," e.g., numbers.

In order to make it clear that I am not trading on an ontological irrelevance of proper names, let me suppose that these have been eliminated in favor of individual descriptions, and let me restate the thesis:

> (C″) Observational reports without proper names can generally be interpreted so that the variables range over entities which would ordinarily be called "theoretical" or "nonobservational," e.g., numbers.

The point is that variables are essentially indices, and the exact nature of the entities that make up the index set is immaterial. Thus, if I count heads, and you count noses, and Smith counts people, we will come up with the same number. To extend the example, suppose that in three different interpretations of a language,

$$\text{'}M\text{' is true of a } \begin{cases} \text{person (in Smith's interpretation)} \\ \text{head (in your interpretation)} \\ \text{nose (in my interpretation)} \end{cases}$$

if and only if

$$\text{the } \begin{cases} \text{person is a man} \\ \text{head is that of a man} \\ \text{nose is that of a man} \end{cases}$$

(It is irrelevant that English, the metalanguage, is slanted toward Smith's interpretation.) A sentence involving only 'M' as a predicate has the same truth value in each of these interpretations. The device can clearly be applied to other predicates (of people, noses, heads, or what have you) to get three interpretations of the calculus which are essentially the same.

Now we can imagine a fourth, numerical interpretation, obtained by assigning numbers to people so that different people have different numbers (e.g., social security numbers). Let S be the set of all such numbers. Then 'M' is true of a number in S in this interpretation if and only if it is the number of a man; and similarly for other predicates. M is then a property of numbers. Suppose that H is the property that applies to a number if and only if the person with that number is the author of "What Theories Are Not." I may have seen someone writing a manuscript entitled "What Theories Are Not," and I may have seen that the person is a man. Then

$$M(\imath x)Hx$$

is an observation report for me at the time in question, even though it is interpreted as attributing a certain property (M) to a certain number (($\imath x)Hx$), that is, to an entity which is an unobservable if anything is.

A moral about the theory of reference can now be illustrated by contrasting the interpretation of Mx as *x is the number of a male* with the interpretation of Mx as *x is an odd number* in the case where in fact all positive integers are social security numbers. It might happen that the males, and only they, have odd social security numbers, in which case the two interpretations are indistinguishable from the point of view of the theory of reference. But in the philosophy of science, one is concerned less with the actual extensions of terms than with the ways in which those extensions are described. Thus, when Braithwaite speaks of a model of a theory as "another interpretation of the theory's calculus," he cannot mean to identify extensionally equivalent interpretations; and in general, some of the most important semantical questions about the functions that assign extensions to the signs of a theory can only be treated by considering those functions in intension. Meaning, like politics, may be imperfectly understood; but this is no ground for ignoring it.

NORWOOD R. HANSON

Observation

Were the eye not attuned to the Sun,
The Sun could never be seen by it.

GOETHE[1]

~~~~~~

A

Consider two microbiologists. They look at a prepared slide; when asked what they see, they may give different answers. One sees in the cell before him a cluster of foreign matter: it is an artefact, a coagulum resulting from inadequate staining techniques. This clot has no more to do with the cell, *in vivo*, than the scars left on it by the archaeologists spade have to do with the original shape of some Grecian urn. The other biologist identifies the clot as a cell organ, a 'Golgi body'. As for techniques, he argues: 'The standard way of detecting a cell organ is by fixing and staining. Why single out this one technique as producing artefacts, while others disclose genuine organs?'

The controversy continues.[2] It involves the whole theory of microscopical technique; nor is it an obviously experimental issue. Yet it affects what scientists say they see. Perhaps there is a sense in which two such observers do not see the same thing, do not begin from the same data, though their eyesight is normal and they are visually aware of the same object.

Imagine these two observing a Protozoon—*Amoeba*. One sees a one-celled animal, the other a non-celled animal. The first sees *Amoeba* in all its analogies with different types of single cells: liver cells, nerve

Reprinted from Chapter I of *Patterns of Discovery*, Norwood R. Hanson, by kind permission of the publishers of the Cambridge University Press.

[1] 'Handeln vom Netz, nicht von dem, was das Netz beschreibt', L. Wittgenstein, *Tractatus Logico-Philosophicus* (Harcourt, Brace and Co., New York, 1922), 6. 35.

> Wär' nicht das Auge sonnenhaft,
> Die Sonne könnt' es nie erblicken;

Goethe, *Zahme Xenien* (Werke, Weimar, 1887–1918), Bk. 3, 1805.

[2] Cf. the papers by Baker and Gatonby in *Nature,* 1949–present.

cells, epithelium cells. These have a wall, nucleus, cytoplasm, and so forth. Within this class *Amoeba* is distinguished only by its independence. The other, however, sees *Amoeba's* homology not with single cells, but with whole animals. Like all animals *Amoeba* ingests its food, digests and assimilates it. It excretes, reproduces and is mobile—more like a complete animal than an individual tissue cell.

This is not an experimental issue, yet it can affect experiment. What either man regards as significant questions or relevant data can be determined by whether he stresses the first or the last term in 'unicellular animal'.[3]

Some philosophers have a formula ready for such situations: 'Of course they see the same thing. They make the same observation since they begin from the same visual data. But they interpret what they see differently. They construe the evidence in different ways.' [4] The task is then to show how these data are moulded by different theories or interpretations or intellectual constructions.

Considerable philosophers have wrestled with this task. But in fact the formula they start from is too simple to allow a grasp of the nature of observation within physics. Perhaps the scientists cited above do not begin their inquiries from the same data, do not make the same observa-

---

[3] This is not a *merely* conceptual matter, of course. Cf. Wittgenstein, *Philosophical Investigations* (Blackwell, Oxford, 1953), p. 196.
[4] (1) G. Berkeley, *Essay Towards a New Theory of Vision* (in *Works*, vol. ɪ (London, T. Nelson, 1948–56), pp. 51 ff.
(2) James Mill, *Analysis of the Phenomena of the Human Mind* (Longmans, London, 1869), vol. ɪ, 97.
(3) J. Sully, *Outlines of Psychology* (Appleton, New York, 1885).
(4) William James, *The Principles of Psychology* (Holt, New York, 1890–1905), vol. ɪɪ, 4, 78, 80 and 81; vol. ɪ, 221.
(5) A. Schopenhauer, *Satz vom Grunde* (in *Sämmtliche Werke*, Leipzig, 1888), ch. ɪv.
(6) H. Spencer, *The Principles of Psychology* (Appleton, New York, 1897), vol. ɪv, chs. ɪx, x.
(7) E. von Hartmann, *Philosophy of the Unconscious* (K. Paul, London, 1931), B, chs. vɪɪ, vɪɪɪ.
(8) W. M. Wundt, *Vorlesungen über die Menschen und Thierseele* (Voss, Hamburg, 1892), ɪv, xɪɪɪ.
(9) H. L. F. von Helmholtz, *Handbuch der Physiologischen Optik* (Leipzig, 1867), pp. 430, 447.
(10) A. Binet, *La psychologie du raisonnement, recherches expérimentales par l'hypnotisme* (Alcan, Paris, 1886), chs. ɪɪɪ, v.
(11) J. Grote, *Exploratio Philosophica* (Cambridge, 1900), vol. ɪɪ, 201 ff.
(12) B. Russell, in *Mind* (1913, p. 76. *Mysticism and Logic* (Longmans, New York, 1918), p. 209. *The Problems of Philosophy* (Holt, New York, 1912), pp. 73, 92, 179, 203.
(13) Dawes Hicks, *Arist. Soc. Sup.* vol. ɪɪ (1919), 176–8.
(14) G. F. Stout, *A Manual of Psychology* (Clive, London, 1907, 2nd ed.), vol. ɪɪ, 1 and 2, 324, 561–4.
(15) A. C. Ewing, *Fundamental Questions of Philosophy* (New York, 1951), pp. 45 ff.
(16) G. W. Cunningham, *Problems of Philosophy* (Holt, New York, 1924), pp. 96–7.

tions, do not even see the same thing? Here many concepts run together. We must proceed carefully, for wherever it makes sense to say that two scientists looking at *x* do not see the same thing, there must always be a prior sense in which they do see the same thing. The issue is, then, 'Which of these senses is most illuminating for the understanding of observational physics?'

These biological examples are too complex. Let us consider Johannes Kepler: imagine him on a hill watching the dawn. With him is Tycho Brahe. Kepler regarded the sun as fixed: it was the earth that moved. But Tycho followed Ptolemy and Aristotle in this much at least: the earth was fixed and all other celestial bodies moved around it. *Do Kepler and Tycho see the same thing in the east at dawn?*

We might think this an experimental or observational question, unlike the questions 'Are there Golgi bodies?' and 'Are Protozoa one-celled or non-celled?' Not so in the sixteenth and seventeenth centuries. Thus Galileo said to the Ptolemaist '. . . neither Aristotle nor you can prove that the earth is *de facto* the centre of the universe . . .' [5] 'Do Kepler and Tycho see the same thing in the east at dawn?' as perhaps not a *de facto* question either, but rather the beginning of an examination of the concepts of seeing and observation.

The resultant discussion might run:
'Yes, they do.'
'No, they don't.'
'Yes, they do!'
'No, they don't!' . . .

That this is possible suggests that there may be reasons for both contentions.[6] Let us consider some points in support of the affirmative answer.

The physical processes involved when Kepler and Tycho watch the dawn are worth noting. Identical photons are emitted from the sun; these traverse solar space, and our atmosphere. The two astronomers have normal vision; hence these photons pass through the cornea, aqueous humour, iris, lens and vitreous body of their eyes in the same way. Finally their retinas are affected. Similar electro-chemical changes occur in their selenium cells. The same configuration is etched on Kepler's retina as on Tycho's. So they see the same thing.

Locke sometimes spoke of seeing in this way: a man sees the sun if his is a normally-formed retinal picture of the sun. Dr. Sir W. Russell Brain speaks of our retinal sensations as indicators and signals. Everything taking place behind the retina is, as he says, 'an intellectual

---

[5] Galileo, *Dialogue Concerning the Two Chief World Systems* (California, 1953); 'The First Day', p. 33.

[6] "Das ist doch kein Sehen!"—"Das ist doch ein Sehen!" Beide müssen sich begrifflich rechtfertigen lassen' (Wittgenstein, *Phil. Inv.* p. 203).

operation based largely on non-visual experience . . .'.[7] What we *see* are the changes in the *tunica retina*. Dr Ida Mann regards the macula of the eye as itself 'seeing details in bright light', and the rods as 'seeing approaching motor-cars'. Dr Agnes Arber speaks of the eye as itself seeing.[8] Often, talk of seeing can direct attention to the retina. Normal people are distinguished from those for whom no retinal pictures can form: we may say of the former that they can see whilst the latter cannot see. Reporting when a certain red dot can be seen may supply the oculist with direct information about the condition of one's retina.[9]

This need not be pursued, however. These writers speak carelessly: seeing the sun is not seeing retinal pictures of the sun. The retinal images which Kepler and Tycho have are four in number, inverted and quite tiny.[10] Astronomers cannot be referring to these when they say they see the sun. If they are hypnotized, drugged, drunk or distracted they may not see the sun, even though their retinas register its image in exactly the same way as usual.

Seeing is an experience. A retinal reaction is only a physical state—a photochemical excitation. Physiologists have not always appreciated the differences between experiences and physical states.[11] People, not their eyes, see. Cameras, and eyeballs, are blind. Attempts to locate within the organs of sight (or within the neurological reticulum behind the eyes) some nameable called 'seeing' may be dismissed. That Kepler and Tycho do, or do not, see the same thing cannot be supported by reference to the physical states of their retinas, optic nerves or visual cortices: there is more to seeing than meets the eyeball.

---

[7] Brain, *Recent Advances in Neurology* (with Strauss) (London, 1929), p. 88. Compare Helmholtz: 'The sensations are signs to our consciousness, and it is the task of our intelligence to learn to understand their meaning' (*Handbuch der Physiologischen Optik* (Leipzig, 1867), vol. III, 433).

See also Husserl, 'Ideen zu einer Reinen Phaenomenologie', in *Jahrbuch für Philosophie*, vol. I (1913), 75, 79, and Wagner's *Handwörtenbuch der Physiologie*, vol. III, section 1 (1846), 183.

[8] Mann, *The Science of Seeing* (London, 1949), pp. 48–49. Arber, *The Mind and the Eye* (Cambridge, 1954). Compare Müller: 'In any field of vision, the retina sees only itself in its spatial extension during a state of affection. It perceives itself as . . . etc.' (*Zur vergleichenden Physiologie des Gesichtesinnes des Menschen und der Thiere* (Leipzig, 1826), p. 54).

[9] Kolin: 'An astigmatic eye when looking at millimeter paper can accommodate to see sharply either the vertical lines or the horizontal lines' (*Physics* (New York, 1950), pp. 570 ff.).

[10] Cf. Whewell, *Philosophy of Discovery* (London, 1860), 'The Paradoxes of Vision'.

[11] Cf. e.g. J. Z. Young, *Doubt and Certainty in Science* (Oxford, 1951, The North Lectures), and Gray Walter's article in *Aspects of Form*, ed. by I. L. Whyte (London, 1953). Compare Newton: 'Do not the Rays of Light in falling upon the bottom of the Eye excite Vibrations in the Tunica Retina? Which Vibrations, being propagated along the solid Fibres of the Nerves into the Brain, cause the Sense of seeing' (*Opticks* (London, 1769), Bk. III, part 1).

Naturally, Tycho and Kepler see the same physical object. They are both visually aware of the sun. If they are put into a dark room and asked to report when they see something—anything at all—they may both report the same object at the same time. Suppose that the only object to be seen is a certain lead cylinder. Both men see the same thing: namely this object—whatever it is. It is just here, however, that the difficulty arises, for while Tycho sees a mere pipe, Kepler will see a telescope, the instrument about which Galileo has written to him.

Unless both are visually aware of the same object there can be nothing of philosophical interest in the question whether or not they see the same thing. Unless they both see the sun in this prior sense our question cannot even strike a spark.

Nonetheless, both Tycho and Kepler have a common visual experience of some sort. This experience perhaps constitutes their seeing the same thing. Indeed, this may be a seeing logically more basic than anything expressed in the pronouncement 'I see the sun' (where each means something different by 'sun'). If what they meant by the word 'sun' were the only clue, then Tycho and Kepler could not be seeing the same thing, even though they were gazing at the same object.

If, however, we ask, not 'Do they see the same thing?' but rather 'What is it that they both see?', an unambiguous answer may be forthcoming. Tycho and Kepler are both aware of a brilliant yellow-white disc in a blue expanse over a green one. Such a 'sense-datum' picture is single and uninverted. To be unaware of it is not to have it. Either it dominates one's visual attention completely or it does not exist.

If Tycho and Kepler are aware of anything visual, it must be of some pattern of colours. What else could it be? We do not touch or hear with our eyes, we only take in light.[12] This private pattern is the same for both observers. Surely if asked to sketch the contents of their visual fields they would both draw a kind of semicircle on a horizon-line.[13] They say they see the sun. But they do not see every side of the sun at once; so what they really see is discoid to begin with. It is but a visual aspect of the sun. In any single observation the sun is a brilliantly luminescent disc, a penny painted with radium.

[12] 'Rot und grün kann ich nur sehen, aber nicht hören' (Wittgenstein, *Phil. Inv.* p. 209).

[13] Cf. 'An appearance is the same whenever the same eye is affected in the same way' (Lambert, *Photometria* (Berlin, 1760)); 'We are justified, when different perceptions offer themselves to us, to infer that the underlying real conditions are different' (Helmholtz, *Wissenschaftliche Abhandlungen* (Leipzig, 1882), vol. II, 656), and Hertz: 'We form for ourselves images or symbols of the external objects; the manner in which we form them is such that the logically necessary (*denknotwendigen*) consequences of the images in thought are invariably the images of materially necessary (*naturnotwendigen*) consequences of the corresponding objects' (*Principles of Mechanics* (London, 1889), p. 1).

Broad and Price make depth a feature of the private visual pattern. However,

So something about their visual experiences at dawn is the same for both: a brilliant yellow-white disc centred between green and blue colour patches. Sketches of what they both see could be identical—congruent. In this sense Tycho and Kepler see the same thing at dawn. The sun appears to them in the same way. The same view, or scene, is presented to them both.

In fact, we often speak in this way. Thus the account of a recent solar eclipse:[14] 'Only a thin crescent remains; white light is now completely obscured; the sky appears a deep blue, almost purple, and the landscape is a monochromatic green . . . there are the flashes of light on the disc's circumference and now the brilliant crescent to the left. . . .' Newton writes in a similar way in the *Opticks*: 'These Arcs at their first appearance were of a violet and blue Colour, and between them were white Arcs of Circles, which . . . became a little tinged in their inward Limbs with red and yellow. . . .'[15] Every physicist employs the language of lines, colour patches, appearances, shadows. In so far as two normal observers use this language of the same event, they begin from the same data: they are making the same observation. Differences between them must arise in the interpretations they put on these data.

Thus, to summarize, saying that Kepler and Tycho see the same thing at dawn just because their eyes are similarly affected is an elementary mistake. There is a difference between a physical state and a visual experience. Suppose, however, that it is argued as above—that

---

Weyl (*Philosophy of Mathematics and Natural Science* (Princeton, 1949), p. 125) notes that a single eye perceives qualities spread out in a *two*-dimensional field, since the latter is dissected by any one-dimensional line running through it. But our conceptual difficulties remain even when Kepler and Tycho keep one eye closed.

Whether or not two observers are having the same visual sense-data reduces directly to the question of whether accurate pictures of the contents of their visual fields are identical, or differ in some detail. We can then discuss the publicly observable pictures which Tycho and Kepler draw of what they see, instead of those private, mysterious entities locked in their visual consciousness. The accurate picture and the sense-datum must be identical; how could they differ?

[14] From the B.B.C. report, 30 June 1954.

[15] Newton, *Opticks*, Bk. II, part 1. The writings of Claudius Ptolemy sometimes read like a phenomenalist's textbook. Cf. e.g. *The Almagest* (Venice, 1515), VI, section 11, 'On the Directions in the Eclipses', 'When it touches the shadow's circle from within', 'When the circles touch each other from without'. Cf. also VII and VIII, IX (section 4). Ptolemy continually seeks to chart and predict 'the appearances' —the points of light on the celestial globe. *The Almagest* abandons any attempt to explain the machinery behind these appearances.

Cf. Pappus: 'The (circle) dividing the milk-white portion which owes its colour to the sun, and the portion which has the ashen colour natural to the moon itself is indistinguishable from a great circle' (*Mathematical Collection* (Hultsch, Berlin and Leipzig, 1864), pp. 554–60).

they see the same thing because they have the same sense-datum experience. Disparities in their accounts arise in *ex post facto* interpretations of what is seen, not in the fundamental visual data. If this is argued, further difficulties soon obtrude.

B

Normal retinas and cameras are impressed similarly by fig. 1.[16] Our visual sense-data will be the same too. If asked to draw what we see, most of us will set out a configuration like fig. 1.

Do we all see the same thing? [17] Some will see a perspex cube viewed from below. Others will see it from above. Still others will see it as a kind of polygonally-cut gem. Some people see only criss-crossed lines in a plane. It may be seen as a block of ice, an aquarium, a wire frame for a kite —or any of a number of other things.

Do we, then, all see the same thing? If we do, how can these differences be accounted for?

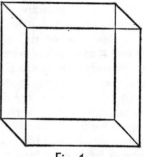

Fig. 1

Here the 'formula' re-enters: 'These are different *interpretations* of what all observers see in common. Retinal reactions to fig. 1 are virtually identical; so too are our visual sense-data, since our drawings of what we see will have the same content. There is no place in the seeing for these differences, so they must lie in the interpretations put on what we see.'

This sounds as if I do two things, not one, when I see boxes and bicycles. Do I put different interpretations on fig. 1 when I see it now as a box from below, and now as a cube from above? I am aware of no such thing. I mean no such thing when I report that the box's perspective has snapped back into the page.[18] If I do not mean this, then the concept of seeing which is natural in this connexion does not designate two diaphanous components, one optical, the other interpretative.

---

[16] This famous illusion dates from 1832, when L. A. Necker, the Swiss naturalist, wrote a letter to Sir David Brewster describing how when certain rhomboidal crystals were viewed on end the perspective could shift in the way now familiar to us. Cf. *Phil. Mag.* III, no. 1 (1832), 329–37, especially 336. It is important to the present argument to note that this observational phenomenon began life not as a psychologist's trick, but at the very frontiers of observational science.

[17] Wittgenstein answers: 'Denn wir sehen eben wirklich zwei verschiedene Tatsachen' (*Tractatus*, 5. 5423).

[18] 'Auf welche Vorgänge spiele ich an?' (Wittgenstein, *Phil. Inv.* p. 214).

Fig. 1 is simply seen now as a box from below, now as a cube from above; one does not first soak up an optical pattern and then clamp an interpretation on it. Kepler and Tycho just see the sun. That is all. That is the way the concept of seeing works in this connexion.

'But', you say, 'seeing fig. 1 first as a box from below, then as a cube from above, involves interpreting the lines differently in each case.' Then for you and me to have a different interpretation of fig. 1 just *is* for us to see something different. This does not mean we see the same thing and then interpret it differently. When I suddenly exclaim 'Eureka—a box from above', I do not refer simply to a different interpretation. (Again, there is a logically prior sense in which seeing fig. 1 as from above and then as from below is seeing the same thing differently, i.e. being aware of the same diagram in different ways. We can refer just to this, but we need not. In this case we do not.)

Besides, the word 'interpretation' is occasionally useful. We know where it applies and where it does not. Thucydides presented the facts objectively. Herodotus put an interpretation on them. The word does not apply to everything—it has a meaning. Can interpreting always be going on when we see? Sometimes, perhaps, as when the hazy outline of an agricultural machine looms up on a foggy morning and, with effort, we finally identify it. Is this the 'interpretation' which is active when bicycles and boxes are clearly seen? Is it active when the perspective of fig. 1 snaps into reverse? There was a time when Herodotus was half-through with his interpretation of the Graeco-Persian wars. Could there be a time when one is half-through interpreting fig. 1 as a box from above, or as anything else?

'But the interpretation takes very little time—it is instantaneous.' Instantaneous interpretation hails from the Limbo that produced unsensed sensibilia, unconscious inference, incorrigible statements, negative facts and *Objektive*. These are ideas which philosophers force on the world to preserve some pet epistemological or metaphysical theory.

Only in contrast to 'Eureka' situations (like perspective reversals, where one cannot interpret the data) is it clear what is meant by saying that though Thucydides could have put an interpretation on history, he did not. Moreover, whether or not an historian is advancing an interpretation is an empirical question: we know what would count as evidence one way or the other. But whether we are employing an interpretation when we see fig. 1 in a certain way is not empirical. What could count as evidence? In no ordinary sense of 'interpret' do I interpret fig. 1 differently when its perspective reverses for me. If there is some extraordinary sense of word it is not clear, either in ordinary language, or in extraordinary (philosophical) language. To insist that different reactions to fig. 1 *must* lie in the interpretations put on a common visual experience is just to reiterate (without reasons) that the seeing of *x* *must* be the same for all observers looking at *x*.

'But "I see the figure as a box" means: I am having a particular visual experience which I always have when I interpret the figure as a box, or when I look at a box. . . .' '. . . if I meant this, I ought to know it. I ought to be able to refer to the experience directly and not only indirectly. . . .' [19]

Ordinary accounts of the experiences appropriate to fig. 1 do not require visual grist going into an intellectual mill: theories and interpretations are 'there' in the seeing from the outset. How can interpretations 'be there' in the seeing? How is it possible to see an object according to an interpretation? 'The question represents it as a queer fact; as if something were being forced into a form it did not really fit. But no squeezing, no forcing took place here.' [20]

Consider now the reversible perspective figures which appear in textbooks on Gestalt psychology: the tea-tray, the shifting (Schröder) staircase, the tunnel. Each of these can be seen as concave, as convex, or as a flat drawing.[21] Do I really see something different each time, or do I only interpret what I see in a different way? To interpret is to think, to do something; seeing is an experiential state.[22] The different ways in which these figures are seen are not due to different thoughts lying behind the visual reactions. What could 'spontaneous' mean if these reactions are not spontaneous? When the staircase 'goes into reverse' it does so spontaneously. One does not think of anything special; one does not think at all. Nor does one interpret. One just sees, now a staircase as from above, now a staircase as from below.

The sun, however, is not an entity with such variable perspective. What has all this to do with suggesting that Tycho and Kepler may see different things in the east at dawn? Certainly the cases are different. But these reversible perspective figures are examples of different things being seen in the same configuration, where this difference is due neither to differing visual pictures, nor to any 'interpretation' superimposed on the sensation.

---

[19] *Ibid.* p. 194 (top).

[20] *Ibid.* p. 200.

[21] This is *not* due to eye movements, or to local retinal fatigue. Cf. Flugel, *Brit. J. Psychol.* vi (1913), 60; *Brit. J. Psychol.* v (1913), 357. Cf. Donahue and Griffiths, *Amer. J. Psychol.* (1931), and Luckiesh, *Visual Illusions and Their Applications* (London, 1922. Cf. also Peirce, *Collected Papers* (Harvard, 1931), 5, 183. References to psychology should not be misunderstood; but as one's acquaintance with the psychology of perception deepens, the character of the conceptual problems one regards as significant will deepen accordingly. Cf. Wittgenstein, *Phil. Inv.* p. 206 (top). Again, p. 193: 'Its causes are of interest to psychologists. We are interested in the concept and its place among the concepts of experience.'

[22] Wittgenstein, *Phil. Inv.* p. 212.

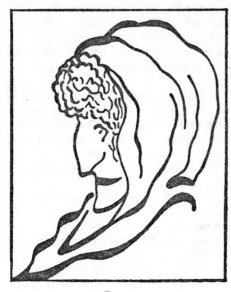

Fig. 2

Some will see in fig. 2 an old Parisienne, others a young woman (à la Toulouse-Lautrec).[23] All normal retinas 'take' the same picture; and our sense-datum pictures must be the same, for even if you see an old lady and I a young lady, the pictures we draw of what we see may turn out to be geometrically indistinguishable. (Some can see this *only* in one way, not both. This is like the difficulty we have after finding a face in a tree-puzzle; we cannot thereafter see the tree without the face.)

When what is observed is characterized so differently as 'young woman' or 'old woman', is it not natural to say that the observers see different things? Or must 'see different things' mean only 'see different objects'? This is a primary sense of the expression, to be sure. But is there not also a sense in which one who cannot see the young lady in fig. 2 sees something different from me, who sees the young lady? Of course there is.

Similarly, in Köhler's famous drawing of the Goblet-and-Faces[24] we 'take' the same retinal/cortical/sense-datum picture of the configuration; our drawings might be indistinguishable. I see a goblet, however, and you see two men staring at one another. Do we see the same thing? Of course we do. But then again we do not. (The sense in which we *do* see the same thing begins to lose its philosophical interest.)

I draw my goblet. You say 'That's just what I saw, two men in a staring contest'. What steps must be taken to get you to see what I see? When attention shifts from the cup to the faces does one's visual picture change? How? What is it that changes? What could change?

[23] From Boring, *Amer. J. Psychol.* XLII (1930), 444 and cf. Allport, *Brit. J. Psychol.* XXI (1930), 133; Leeper, *J. Genet. Psychol.* XLVI (1935), 41: Street, *Gestalt Completion Test* (Columbia Univ., 1931); Dees and Grindley, *Brit. J. Psychol.* (1947).

[24] Köhler, *Gestalt Psychology* (London, 1929). Cf. his *Dynamics in Psychology* (London, 1939).

Nothing optical or sensational is modified. Yet one sees different things. The organization of what one sees changes.[25]

How does one describe the difference between the *jeune fille* and the *vieille femme* in fig. 2? Perhaps the difference is not describable: it may just show itself.[26] That two observers have not seen the same things in fig. 2 could show itself in their behaviour. What is the difference between us when you see the zebra as black with white stripes and I see it as white with black stripes? Nothing optical. Yet there might be a context (for instance, in the genetics of animal pigmentation), where such a difference could be important.

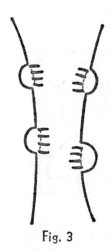

A third group of figures will stress further this organizational element of seeing and observing. They will hint at how much more is involved when Tycho and Kepler witness the dawn that 'the formula' suggests.

What is portrayed in fig. 3? Your retinas and visual cortices are affected much as mine are; our sense-datum pictures would not differ. Surely we could all produce an accurate sketch of fig. 3. Do we see the same thing?

I see a bear climbing up the other side of a tree. Did the elements 'pull together'/cohere/organize, when you learned this? [27] You might even say with Wittgenstein 'it has not changed, and yet I see it differently . . .'.[28] Now, does it not have '. . . a quite particular "organization" '?

Fig. 3

[25] 'Mein Gesichtseindruck hat sich geändert;—wie war er früher; wie ist er jetzt? —Stelle ich ihn durch eine genaue Kopie dar—und ist das keine gute Darstellung? —so zeigt sich keine Änderung' (Wittgenstein, *Phil. Inv.* p. 196).

[26] 'Was gezeigt werden kann, kann nicht gesagt werden' (Wittgenstein, *Tractatus,* 4. 1212).

[27] This case is different from fig. 1. Now I can help a 'slow' percipient by tracing in the outline of the bear. In fig. 1 a percipient either gets the perspectival arrangement, or he does not, though even here Wittgenstein makes some suggestions as to how one might help; cf. *Tractatus*, 5. 5423, last line.

[28] Wittgenstein, *Phil. Inv.* p. 193. Helmholtz speaks of the 'integrating' function which converts the figure into the appearance of an object hit by a visual ray (*Phys. Optik*, vol. III, p. 239). This is reminiscent of Aristotle, for whom seeing consisted in emanations from our eyes. They reach out, tentacle-fashion, and touch objects whose shapes are 'felt' in the eye. (Cf. *De Caelo* (Oxford, 1928), 290 a, 18; and *Meteorologica* (Oxford, 1928), III, iv, 373 b, 2.) (Also Plato, *Meno*, London, 1869, 76 c–d.) But he controverts this in *Topica* (Oxford, 1928), 105 b, 6.) Theophrastus argues that 'Vision is due to the gleaming . . . which [in the eye] reflects to the object' (*On the Senses*, 26, trans. G. M. Stratton). Hero writes: 'Rays proceeding from our eyes are reflected by mirrors . . . that our sight is directed in straight lines proceeding from the organ of vision may be substantiated as follows' (*Ca-*

Organization is not itself seen as are the lines and colours of a drawing. It is not itself a line, shape, or a colour. It is not an element in the visual field, but rather the way in which elements are appreciated. Again, the plot is not another detail in the story. Nor is the tune just one more note. Yet without plots and tunes details and notes would not hang together. Similarly the organization of fig. 3 is nothing that registers on the retina along with other details. Yet it gives the lines and shapes a pattern. Were this lacking we would be left with nothing but an unintelligible configuration of lines.

How do visual experiences become organized? How is seeing possible?

Consider fig. 4 in the context of fig. 5:

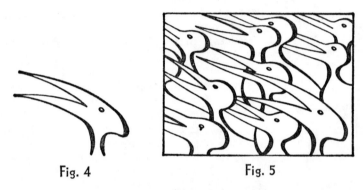

Fig. 4   **Fig. 5**

The context gives us the clue. Here, some people could not see the figure as an antelope. Could people who had never seen an antelope, but only birds, see an antelope in fig. 4?

In the context of fig. 6 the figure may indeed stand out as an antelope. It might even be urged that the figure seen in fig. 5 has no similarity to the one in fig. 6 although the two are congruent. Could anything be more opposed to a sense-datum account of seeing?

Of a figure similar to the Necker cube (fig. 1) Wittgenstein writes,

---

toptrics, 1–5, trans. Schmidt in *Heronis Alexandrini Opera* (Leipzig, 1899–1919)). Galen is of the same opinion. So too is Leonardo: 'The eye sends its image to the object . . . the power of vision extends by means of the visual rays . . .' (*Note-books,* c.a. 135 v.b. and 270 v.c.). Similarly Donne in *The Ecstasy* writes: 'Our eye-beams twisted and . . . pictures in our eyes to get was all *our* propagation.'

This is the view that all perception is really a species of touching, e.g. Descartes' *impressions,* and the analogy of the wax. Compare: '[Democritus] explains [vision] by the air between the eye and the object [being] compressed . . . [it] thus becomes imprinted . . . "as if one were to take a mould in wax" . . .' Theophrastus (*op. cit.* 50–53). Though it lacks physical and physiological support, the view is attractive in cases where lines seem suddenly to be forced into an intelligible pattern—by us.

'You could imagine [this] appearing in several places in a text-book. In the relevant text something different is in question every time: here a glass cube, there an inverted open box, there a wire frame of that

Fig. 6

shape, there three boards forming a solid angle. Each time the text supplies the interpretation of the illustration. But we can also see the illustration now as one thing, now as another. So we interpret it, and see it as we interpret it.' [29]

Consider now the head-and-shoulders in fig. 7:

Fig. 7

The upper margin of the picture cuts the brow, thus the top of the head is not shown. The point of the jaw, clean shaven and brightly illuminated, is just above the geometric center of the picture. A white mantle covers the right shoulder. The right upper sleeve is exposed as

[29] *Ibid.* p. 193. Cf. Helmholtz, *Phys. Optik,* vol. III, 4, 18 and Fichte (*Bestimmung des Menschen,* ed. Medicus (Bonn, 1834), vol. III, 326). Cf. also Wittgenstein, *Tractatus,* 2. 0123.

the rather black area at the lower left. The hair and beard are after the manner of a late mediaeval representation of Christ.[30]

The appropriate aspect of the illustration is brought out by the verbal context in which it appears. It is not an illustration of anything determinate unless it appears in some such context. In the same way, I must talk and gesture around fig. 4 to get you to see the antelope when only the bird has revealed itself. I must provide a context. The context is part of the illustration itself.

Such a context, however, need not be set out explicitly. Often it is 'built into' thinking, imagining and picturing. We are set[31] to appreciate the visual aspect of things in certain ways. Elements in our experience do not cluster at random.

A trained physicist could see one thing in fig. 8: an X-ray tube

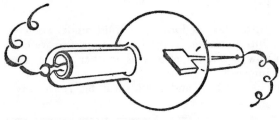

Fig. 8

viewed from the cathode. Would Sir Lawrence Bragg and an Eskimo baby see the same thing when looking at an X-ray tube? Yes, and no. Yes—they are visually aware of the same object. No—the *ways* in which they are visually aware are profoundly different. Seeing is not only the having of a visual experience; it is also the way in which the visual experience is had.

---

[30] P. B. Porter, *Amer. J. Psychol.* LXVII (1954), 550.

[31] Writings by Gestalt psychologists on 'set' and 'Aufgabe' are many. Yet they are overlooked by most philosophers. A few fundamental papers are: Külpe, *Ber. I Kongress Exp. Psychol., Giessen* (1904); Bartlett, *Brit. J. Psychol.* VIII (1916), 222; George, *Amer. J. Psychol.* XXVIII (1917), 1; Fernberger, *Psychol. Monogr.* XXVI (1919), 6; Zigler, *Amer. J. Psychol.* XXXI (1920), 273; Boring, *Amer. J. Psychol.* XXXV (1924), 301; Wilcocks, *Amer. J. Psychol.* XXXVI (1925), 324; Gilliland, *Psychol. Bull.* XXIV (1927), 622; Gottschaldt, *Psychol. Forsch.* XII (1929), 1; Boring, *Amer. J. Psychol.* XLII (1930), 444; Street, *Gestalt Completion Test* (Columbia University, 1931); Ross and Schilder, *J. Gen. Psychol.* X (1934), 152; Hunt, *Amer. J. Psychol.* XLVII (1935), 1; Süpola, *Psychol. Monogr.* XLVI (1935), 210, 27; Gibson, *Psychol. Bull.* XXXVIII (1941), 781; Henle, *J. Exp. Psychol.* XXX (1942), 1; Luchins, *J. Soc. Psychol.* XXI (1945), 257; Wertheimer, *Productive Thinking* (1945); Russell Davis and Sinha, *Quart. J. Exp. Psychol.* (1950); Hall, *Quart. J. Exp. Psychol.* II (1950), 153.

Philosophy has no concern with fact, only with conceptual matters (cf. Wittgenstein, *Tractatus*, 4. 111); but discussions of perception could not be improved by the reading of these twenty papers.

At school the physicist had gazed at this glass-and-metal instrument. Returning now, after years in University and research, his eye lights upon the same object once again. Does he see the same thing now as he did then? Now he sees the instrument in terms of electrical circuit theory, thermodynamic theory, the theories of metal and glass structure, thermionic emission, optical transmission, refraction, diffraction, atomic theory, quantum theory and special relativity.

Contrast the freshman's view of college with that of his ancient tutor. Compare a man's first glance at the motor of his car with a similar glance ten exasperating years later.

'Granted, one learns all these things', it may be countered, 'but it all figures in the interpretation the physicist puts on what he sees. Though the layman sees exactly what the physicist sees, he cannot interpret it in the same way because he has not learned so much.'

Is the physicist doing more than just seeing? No; he does nothing over and above what the layman does when he sees an X-ray tube. What are you doing over and above reading these words? Are you interpreting marks on a page? When would this ever be a natural way of speaking? Would an infant see what you see here, when you see words and sentences and he sees but marks and lines? One does nothing beyond looking and seeing when one dodges bicycles, glances at a friend, or notices a cat in the garden.

'The physicist and the layman see the same thing', it is objected, 'but they do not make the same thing of it.' The layman can make nothing of it. Nor is that just a figure of speech. I can make nothing of the Arab word for *cat*, though my purely visual impressions may be indistinguishable from those of the Arab who can. I must learn Arabic before I can see what he sees. The layman must learn physics before he can see what the physicist sees.

If one must find a paradigm case of seeing it would be better to regard as such not the visual apprehension of colour patches but things like seeing what time it is, seeing what key a piece of music is written in, and seeing whether a wound is septic.[32]

Pierre Duhem writes:

> Enter a laboratory; approach the table crowded with an assortment of apparatus, an electric cell, silk-covered copper wire, small cups of mercury, spools, a mirror mounted on an iron bar; the experimenter is inserting into small openings the metal ends of ebony-headed pins; the iron oscillates, and the mirror attached to it throws a luminous band upon a celluloid scale; the forward-backward motion of this spot enables the physicist to observe the minute oscillations of the iron bar. But ask him what he is doing. Will he answer 'I am studying the oscillations of

[32] Often 'What do you see?' only poses the question 'Can you identify the object before you?'. This is calculated more to test one's knowledge than one's eyesight.

an iron bar which carries a mirror'? No, he will say that he is measuring the electric resistance of the spools. If you are astonished, if you ask him what his words mean, what relation they have with the phenomena he has been observing and which you have noted at the same time as he, he will answer that your question requires a long explanation and that you should take a course in electricity.[33]

The visitor must learn some physics before he can see what the physicist sees. Only then will the context throw into relief those features of the objects before him which the physicist sees as indicating resistance.

This obtains in all seeing. Attention is rarely directed to the space between the leaves of a tree, save when a Keats brings it to our notice.[34] (Consider also what was involved in Crusoe's seeing a vacant space in the sand as a footprint.) Our attention most naturally rests on objects and events which dominate the visual field. What a blooming, buzzing, undifferentiated confusion visual life would be if we all arose tomorrow without attention capable of dwelling only on what had heretofore been overlooked.[35]

The infant and the layman can see: they are not blind. But they cannot see what the physicist sees; they are blind to what he sees.[36] We may not hear that the oboe is out of tune, though this will be painfully obvious to the trained musician. (Who, incidentally, will not hear the tones and *interpret* them as being out of tune, but will simply hear the oboe to be out of tune.[37] We simply see what time it is; the surgeon simply sees a wound to be septic; the physicist sees the X-ray tube's anode overheating.) The elements of the visitor's visual field, though identical with those of the physicist, are not organized for him as for the physicist; the same lines, colours, shapes are apprehended by both, but not in the same way. There are indefinitely many ways in which a constellation of lines, shapes, patches, may be seen. *Why* a

[33] Duhem, *La théorie physique* (Paris, 1914), p. 218.

[34] Chinese poets felt the significance of 'negative features' like the hollow of a clay vessel or the central vacancy of the hub of a wheel (cf. Waley, *Three Ways of Thought in Ancient China* (London, 1939), p. 155).

[35] Infants are indiscriminate; they take in spaces, relations, objects and events as being of equal value. They still must learn to organize their visual attention. The camera-clarity of their visual reactions is not by itself sufficient to differentiate elements in their visual fields. Contrast Mr. W. H. Auden who recently said of the poet that he is 'bombarded by a stream of varied sensations which would drive him mad if he took them all in. It is impossible to guess how much energy we have to spend every day in not-seeing, not-hearing, not-smelling, not-reacting.'

[36] Cf. 'He was blind to the *expression* of a face. Would his eyesight on that account be defective?' (Wittgenstein, *Phil. Inv.* p. 210) and 'Because they seeing see not; and hearing they hear not, neither do they understand' (Matt. xiii. 10–13).

[37] 'Es hört doch jeder nur, was er versteht' (Goethe, *Maxims* (*Werke*, Weimar, 1887–1918)).

visual pattern is seen differently is a question for psychology, but *that* it may be seen differently is important in any examination of the concepts of seeing and observation. Here, as Wittgenstein might have said, the psychological is a symbol of the logical.

You see a bird, I see an antelope; the physicist sees an X-ray tube, the child a complicated lamp bulb; the microscopist sees coelenterate mesoglea, his new student sees only a gooey, formless stuff. Tycho and Simplicius see a mobile sun, Kepler and Galileo see a static sun.[38]

It may be objected, 'Everyone, whatever his state of knowledge, will see fig. 1 as a box or cube, viewed as from above or as from below'. True; almost everyone, child, layman, physicist, will see the figure as box-like one way or another. But could such observations be made by people ignorant of the construction of box-like objects? No. This objection only shows that most of us—the blind, babies, and dimwits excluded—have learned enough to be able to see this figure as a three-dimensional box. This reveals something about the sense in which Simplicius and Galileo do see the same thing (which I have never denied): they both see a brilliant heavenly body. The schoolboy and the physicist both see that the X-ray tube will smash if dropped. Examining how observers see different things in *x* marks something important about their seeing the same thing when looking at *x*. If seeing different things involves having different knowledge and theories about *x*, then perhaps the sense in which they see the same thing involves their sharing knowledge and theories about *x*. Bragg and the baby share no knowledge of X-ray tubes. They see the same thing only in that if they are looking at it they are both having some visual experience of it. Kepler and Tycho agree on more: they see the same thing in a stronger sense. Their visual fields are organized in much the same way. Neither sees the sun about to break out in a grin, or about to crack into ice cubes. (The baby is not 'set' even against these eventualities.) Most people today see the same thing at dawn in an even stronger sense: we share much knowledge of the sun. Hence Tycho and Kepler see different things, and yet they see the same thing. That these things can be said depends on their knowledge, experience, and theories.

Kepler and Tycho are to the sun as we are to fig. 4, when I see the

---

[38] Against this Professor H. H. Price has argued: 'Surely it appears to both of them to be rising, to be moving upwards, across the horizon . . . they both see a moving sun: they both see a round bright body which appears to be rising.' Philip Frank retorts: 'Our sense observation shows only that in the morning the distance between horizon and sun is increasing, but it does not tell us whether the sun is ascending or the horizon is descending . . .' (*Modern Science and Its Philosophy* (Harvard, 1949), p. 231). Precisely. For Galileo and Kepler the horizon drops; for Simplicius and Tycho the sun rises. This is the difference Price misses, and which is central to this essay.

bird and you see only the antelope. The elements of their experiences are identical; but their conceptual organization is vastly different. Can their visual fields have a different organization? Then they can see different things in the east at dawn.

It is the sense in which Tycho and Kepler do not observe the same thing which must be grasped if one is to understand disagreements within microphysics. Fundamental physics is primarily a search for intelligibility—it is philosophy of matter. Only secondarily is it a search for objects and facts (though the two endeavours are as hand and glove). Microphysicists seek new modes of conceptual organization. If that can be done the finding of new entities will follow. Gold is rarely discovered by one who has not got the lay of the land.

To say that Tycho and Kepler, Simplicius and Galileo, Hooke and Newton, Priestley and Lavoisier, Soddy and Einstein, De Broglie and Born, Heisenberg and Bohm all make the same observations but use them differently is too easy.[39] It does not explain controversy in research science. Were there no sense in which they were different observations they could not be used differently. This may perplex some: that researchers sometimes do not appreciate data in the same way is a serious matter. It is important to realize, however, that sorting out differences about data, evidence, observation, may require more than simply gesturing at observable objects. It may require a comprehensive reappraisal of one's subject matter. This may be difficult, but it should not obscure the fact that nothing less than this may do.

c

There is a sense, then, in which seeing is a 'theory-laden' undertaking. Observation of x is shaped by prior knowledge of x. Another influence on observations rests in the language or notation used to express what we know, and without which there would be little we could recognize as knowledge.

---

[39] This parallels the too-easy epistemological doctrine that all normal observers see the same things in x, but interpret them differently.

PAUL K. FEYERABEND

On the Interpretation
of Scientific Theories

I

According to *positivism* the interpretation of scientific theories is a
function of either experience, or of some observational language.
There are various views about the nature of this function and we may
therefore distinguish a great many varieties of positivism, namely
(*a*) theoretical terms are explicitly definable on the basis of observational terms; (*b*) theoretical terms are extensionally reducible to observational terms; (*c*) theoretical terms are intensionally reducible to
observational terms; (*d*) theoretical terms are implicitly definable with
the help of interpretative systems which (1) do not, (2) do contain
probability statements; and so on. In the present paper I shall discuss
two objections which may be raised against all these varieties. Indeed,
it seems to me that the difficulties of positivism as revealed by these
objections cannot be overcome by inventing a new, and ingenious
connexion between theoretical terms and observational terms, but
only by altogether dropping the idea that the meaning of theoretical
terms depends upon such a connexion.

II

My *first objection* against positivism is that it implies that statements
describing causally independent situations may yet be semantically
dependent. In order to understand this objection consider the attempt
to explicate statements about material objects in terms of what is seen,
heard, felt by observers of a certain kind. It is well known that what
is seen, heard, felt by observers depends upon the object as well as
upon the physiological status of the observers themselves. This status

Reprinted by kind permission of the author and publishers from the *Proceedings of the 12th International Congress of Philosophy*, Venice, 1958, Vol. 5, 151–59.

may be changed by influences (drugs, hypnosis etc.) which act independently of the influences of the observed objects. Any attempt to explain the properties of material objects on the basis of experience will have to take such additional influences into account. An explication will therefore be of the form

$$F(M; S; O) \qquad (1)$$

where F is a complicated logical constant, not necessarily extensional; M a (general or particular) situation pertaining to material objects; O a (general or particular) observable situation; and S a situation which we shall call the mediating situation and which is identical with the conditions of observation. In a special case these conditions may include a reference to the intensity of the light irradiating the object; to the absence of obstacles between the object and the observer at the time of observation; to the properties of the retina and of the brain of the observer and to many other situations which, although *causally independent* of the observed object (turning out the light does not influence the material object, although it makes it invisible), do yet contribute to its observable effect. Those terms of "S" which are neither observational terms, nor terms of "M", will be called the *mediating terms*. In our above example the mediating terms are "light"; "intervening obstacle"; "retina"; "brain"; and others. Now if we adopt the principle, characteristic for the positivistic approach, that the descriptive terms of "M", "obtain only an indirect, and incomplete interpretation by the fact that some of them are connected . . . with observational terms" [1], then we shall have to assume that their interpretation is implicitly defined by statement (1), and therefore *dependent upon all terms of "S"*, although we know at the same time that the situation M is causally dependent only upon part of the situation S, and possibly altogether independent of it.

As a second example consider the attempt to explain the theoretical terms of celestial mechanics on the basis of observational terms referring to bright dots as seen either through a telescope, or upon a photographic plate. In this case the mediating situation consists in the optical properties of the planets, the properties of the light which is reflected by them, the properties of the atmosphere of the earth, the properties of telescopes, and so on. Again the interpretation of sentences containing the terms to be explicated will depend upon the interpretation of other sentences referring to states of affairs which are in no causal relation whatever to the states of affairs referred to by the former. For example, the interpretation (the "meaning") of "mass of the sun" will partly depend upon the interpretation of "refractive index of the atmosphere of the earth".

---

[1] R. Carnap, *The Methodological Character of Theoretical Concepts,* in "Minnesota Studies in the Philosophy of Science", Vol. I, 47.

We shall now discuss three methods which one might adopt in order to overcome this difficulty. The first method consists in denying that situations described with the help of theoretical terms only either exist, or can be regarded as causes, or elements of other situations not so describable. This move is hardly appropriate for a philosopher who has set himself the task to analyse physics, as it completely disregards the existential character of general scientific theories.[2] The second method consists in the attempt to eliminate the mediating terms from the rules of correspondence which connect theoretical terms and observational terms. It is obvious that simple omission of the mediating terms will not do. For a necessary criterion of the adequacy of any logical reconstruction of science is that it translates true sentences into true sentences, and it can easily be shown that the statements resulting from the original rules of correspondence by omission of the mediating terms will in general be empirically false ("table in x at time $t =$ anybody inspecting region x at time t perceives a table" is empirically false; in a dark room no table will be seen by anybody). But the attempt to give an observational account of the mediating terms cannot succeed either, as any such account involves further mediating terms and will therefore never come to an end. The third method which has not yet led to any concrete suggestion would have to consist in devising semantical rules which make the interpretation of the theoretical terms dependent upon the interpretation of the observational terms only, and this in spite of the fact, that the rules of correspondence contain the mediating terms as well. It is difficult to see how such a procedure would account for differences in the interpretation of theoretical terms without dropping the principle that they do not possess an independent meaning: we may safely assume that "Ax" means something different in "if a colourblind man inspects x and sees grey, then Ax"; and in "if a man with perfect coloursight inspects x and sees grey, then Ax", and yet the method discussed at the present moment does not allow us to explain this difference by pointing to a difference in the observational terms employed. It is also difficult to see how a change of the logical constant F in formula (1) could do the trick. For although such a change may influence the *kind* of dependence existing between S and M, it will not, and cannot (see our discussion of the third method, immediately above) *eliminate* this influence.

III

My *second objection* is closely related to the first. It may be stated by saying that given two situations, S' and S'' which are causally independent of each other, a change of our theories about S' which pre-

[2] Cf. H. Feigl, *Existential Hypotheses*, "Phil. of Science", xvii, 35–62.

serves this causal independence may yet imply a change of the interpretation of S''. I shall explain this second objection with the help of
an example consisting of a formalism T expressing the state of affairs
T', which is interpreted by an interpretative system J.

Example: T is a formalization of celestial mechanics T'. The descriptive terms of that formalism are functors, such as "mass", "force",
"acceleration", "heliocentric coordinate", and so forth, whose variables
range over particles of matter. I shall assume that the observation
terms are again functors, such as "declination", "high ascension" whose
variables range over luminescent points in the sky. The interpretative
system used will be fairly complicated as it must take into account
refraction, aberration and so forth. Now the behavior of the planets
and, more especially, the properties of the force acting between them
may be safely said to be causally independent of the thermal and optical properties of the atmosphere of the earth. Yet, if the "meaning"
of the theoretical terms is to depend upon their connexion with observation and upon nothing else (see the quotation from Carnap in
the text to footnote 1), then any change of our assumptions about these
latter properties will necessitate a change of the interpretative system
used, and thereby effect the interpretation of the theoretical terms of
T.

To sum up: the first objection is that according to positivism statements describing causally independent situations will yet be semantically dependent; the second objection is that a change of our *knowledge* concerning a situation S^1 which is causally independent of
another situation S^2 will yet imply a change in the interpretation of the
terms of "S^2". Both objections are based upon the principle (which I
shall call the *principle of semantic independence*) that the interpretation of a statement describing a situation which is causally independent of the situation described by another statement, should be independent of the interpretation of this latter statement.

IV

The above objections cannot be removed by pointing out that the
logical constant F in formula (1) can be chosen in such a way that a
complete interpretation of theoretical terms is never obtained. For our
quarrel is not with the fact that, given a certain interpretative system
the positivistic method of interpretation does not leave room for a
further specification of meaning: it is with the fact that *any* specification, however incomplete and 'open', which is based upon formula
(1) contradicts the principle of semantic independence and must therefore be regarded inadequate.

Another attempt to escape our two objections consists in advocating
probabilistic rules of correspondence. This move which has been suggested by A. Pap seems to be inspired by the realization that a more

'liberal' account of the connexion between the meaning of theoretical terms and the meaning of observational terms is required. Pap seems to agree[3] with my criticism as far as non probabilistic F's are concerned. He seems to assume, however, that a probabilistic F will solve the difficulties. My criticism of Pap's proposal will proceed in two steps. I consider first his assertion that different probabilistic interpretative systems need not contradict each other (whereas different non probabilistic interpretative systems for the same theory will frequently contradict each other). Now this is a point which I would hardly contest and I do not see how it can be regarded as a solution of the difficulties I have pointed out. For I have not attacked the positivistic theory of meaning specification on the grounds that two *inconsistent* interpretative systems lead to different meanings of the theoretical terms, but rather on the grounds that *any two different* interpretative systems lead to different meanings of the theoretical terms, and this in spite of the fact that the difference may be one of situations which are causally independent of whatever theoretical situation may obtain. Hence, it is sufficient for me that $P(T/A) = p'$ and $P(T/A\&B) = p''$ are different, the second implying that T depends upon B, the first not implying any such assertion.*

This last example leads at once to our decisive second attack against probabilistic meaning specification. For assume T to be a state of affairs described in theoretical terms only, O an observational situation, and M the mediating situation. Now if, as is frequently the case, T and M are causally independent, then $P(T/M\&O) = P(T/O)$, that is mediating situations, *although necessary as conditions of observation* (cf. our objections against the second method discussed in sec. 2) *can be eliminated from any probabilistic statement about the relation of theoretical terms to observational terms.* This quite obviously shows that *a probabilistic F is not only inadequate as a meaning rule, but also as a statement expressing which tests are relevant and which are not.* This finishes the probabilistic rescue manoeuvre.

V

It is instructive to trace the origin of the difficulty of positivism as expressed in our two objections. According to physical theory any observable state of affairs, that is any state of affairs which is big enough to be accessible to inspection by human beings, is the result of a superposition of many influences. According to the positivistic account of these theories every single influence contributing to the observational state of affairs is to be described with the help of theo-

---

[3] Private communication.

* $P(T/C) = p$ is the symbolization of the assertion that the probability of T given that C is true is equal to p. —Ed.

retical terms. Briefly, and crudely we may therefore say that according to *physics* any observational situation is the result of a superposition of many theoretical entities which are partly dependent upon, partly independent of each other; and it is therefore a complicated and intricate affair. *Positivism* turns the situation upside down.[4] For a positivist the observational situations are the primitive and unanalysable elements in terms of which the theories must be understood. Now if we realize that the theoretical entities of a given theory make *only a very small contribution* to any observational situation, then the attempt to explain them *exclusively* on the basis of observation will at once be recognized as absurd. It is absurd as it regards as simple what is complicated, and as it attempts to explain a state of affairs (the state some theoretical entity) in terms of other states of affairs to which it makes only a sometimes very insignificant contribution. It may be pointed out that for Aristotelian physics this absurdity does not exist. For the Aristotelians were much more inclined to take observable states of affairs at their face value. But it is impossible to believe at the same time in physics *and* in the above account of the interpretation of its terms.

VI

Having criticized the positivistic theory of the interpretation of general empirical terms, we now turn to positive suggestions. The result of our criticism was that the interpretation of scientific terms must be independent of their occurrence in statements of the form (1), or, what amounts to the same, it must be independent of their connexion with experience. As a theory is testable only insofar as it is connected with experience *via* these very statements, it follows that if a theory is to be meaningful at all, its interpretation must go beyond whatever counts as its 'empirical content': *the interpretation of any physical theory contains metaphysical elements,* the terms 'metaphysical' here being used as synonymous with 'nonempirical'. If this statement seems surprising it is partly because of the new, and 'technical' sense which the word 'metaphysical' has received in the hands of the positivists. In the next section we shall show that this result which we have here derived on the basis of the principle of semantic independence, has a natural place within *realism*.

VII

Realism asserts that there exist states of affairs which are causally independent of the states of observers, measuring instruments and the like, but which may influence these instruments and these ob-

---

[4] This has been pointed out by Prof. H. Feigl in his *Existential Hypotheses,* whose point of view is in many respects similar to the one defended here, although it may not be quite as radical.

servers. It also admits that whatever influence a real state of affairs exerts upon an observer, it will not be the only influence, but will have to interact with many other influences, some of them known, some of them unknown. According to the realistic interpretation a scientific theory aims at a description of states of affairs, or properties of physical systems which transcends experience not only insofar as it is general (whereas any description of experience can only be singular), but also insofar as it *disregards all the independent causes which apart from the situations described by the theory may influence the observer or his measuring instrument.* For example, Newtonian astronomy describes the structure of the planetary system, the mutual interaction of the planets and their behaviour, without taking into account all the disturbances experienced by the light which, having left the sun, having been reflected and diffracted by the atmosphere of the earth as well as in the lens of some telescope, reports this structure only in a more or less distorted way. Of course, any attempt to *test* Newtonian astronomy will have to take these distortions into account; for a test of a situation consists in connecting a cause *C* provided by it with an effect to which also other causes have contributed; and it presupposes that all these causes are known as well. But this must of course not be taken to imply that the "meaning" of the statement that *C* obtains, depends upon the meaning of statements describing those other causes. *The interpretation of a scientific theory depends upon nothing but the state of affairs it describes.*[5] This is an immediate consequence of the principle of semantic independence.

This result has consequences with respect to the interpretation of metaphysics. A metaphysical theory does not contain any indication as to how we are to test it, a scientific theory contains some such indications without thereby making its *whole* content accessible to test. On the basis of what we said above it is no longer possible to distinguish "metaphysical meaning" (or "nonsense") and "scientific meaning" by referring to testability (although testability may of course be regarded as a useful criterion for separating scientific *theories* from metaphysical *theories*). For it may happen that a scientific theory and a metaphysical theory describe exactly the same state of affairs, the one in a testable way, the other in a way which is inaccessible to test (in a similar way to that in which a sentence which is decidable in one theory, and undecidable in another may yet in both cases refer to the same state of affairs) and that they therefore possess identically the same meaning. But the discussion of this possibility would already transcend the scope of the present paper which was to criticize positivism.

University of California, Berkeley

[5] For this cf. also Feigl, *Existential Hypotheses.*

W. V. O. QUINE

Posits and Reality

1. SUBVISIBLE PARTICLES

According to physics my desk is, for all its seeming fixity and solidity, a swarm of vibrating molecules. The desk as we sense it is comparable to a distant haystack in which we cannot distinguish the individual stalks; comparable also to a wheel in which, because of its rapid rotation, we cannot distinguish the individual spokes. Comparable, but with a difference. By approaching the haystack we can distinguish the stalks, and by retarding the wheel we can distinguish the spokes. On the other hand no glimpse is to be had of the separate molecules of the desk; they are, we are told, too small.

Lacking such experience, what evidence can the physicist muster for his doctrine of molecules? His answer is that there is a convergence of indirect evidence, drawn from such varied phenomena as expansion, heat conduction, capillary attraction, and surface tension. The point is that these miscellaneous phenomena can, if we assume the molecular theory, be marshaled under the familiar laws of motion. The fancifulness of thus assuming a substructure of moving particles of imperceptible size is offset by a gain in naturalness and scope on the part of the aggregate laws of physics. The molecular theory is felt, moreover, to gain corroboration progressively as the physicist's predictions of future observations turn out to be fulfilled, and as the

Written about 1955 for the beginning of *Word and Object,* but eventually superseded. First published along with a Japanese translation in S. Uyeda, ed. *Basis of the Contemporary Philosophy*, Vol. 5 (Tokyo: Waseda University Press, 1960). It has appeared also in Italian, *Rivista di Filosofia,* 1964.

Reprinted by kind permission of the author and publishers from S. Uyeda, ed., *Basis of the Contemporary Philosophy,* Vol. 5 (Tokyo: Waseda University Press, 1960).

theory proves to invite extensions covering additional classes of phenomena.

The benefits thus credited to the molecular doctrine may be divided into five. One is simplicity: empirical laws concerning seemingly dissimilar phenomena are integrated into a compact and unitary theory. Another is familiarity of principle: the already familiar laws of motion are made to serve where independent laws would otherwise have been needed. A third is scope: the resulting unitary theory implies a wider array of testable consequences than any likely accumulation of separate laws would have implied. A fourth is fecundity: successful further extensions of theory are expedited. The fifth goes without saying: such testable consequences of the theory as have been tested have turned out well, aside from such sparse exceptions as may in good conscience be chalked up to unexplained interferences.

Simplicity, the first of the listed benefits, is a vague business. We may be fairly sure of this much: theories are more or less simple, more or less unitary, only relative to one or another given vocabulary or conceptual apparatus. Simplicity is, if not quite subjective, at any rate parochial. Yet simplicity contributes to scope, as follows. An empirical theory, typically, generalizes or extrapolates from sample data, and thus covers more phenomena than have been checked. Simplicity, by our lights, is what guides our extrapolation. Hence the simpler the theory, on the whole, the wider this unchecked coverage.

As for the fourth benefit, fecundity, obviously it is a consequence of the first two, simplicity and familiarity, for these two traits are the best conditions for effective thinking.

Not all the listed benefits are generally attributable to accepted scientific theories, though all are to be prized when available. Thus the benefit of familiarity of principle may, as in quantum theory and relativity theory, be renounced, its loss being regretted but outweighed.

But to get back. In its manifest content the molecular doctrine bears directly on unobservable reality, affirming a structure of minute swarming particles. On the other hand any defense of it has to do rather with its indirect bearing on observable reality. The doctrine has this indirect bearing by being the core of an integrated physical theory which implies truths about expansion, conduction, and so on. The benefits which we have been surveying are benefits which the molecular doctrine, as core, brings to the physics of these latter observable phenomena.

Suppose now we were to excise that core but retain the surrounding ring of derivative laws, thus not disturbing the observable consequences. The retained laws could be viewed thenceforward as autonomous empirical laws, innocent of any molecular commitment. Granted, this combination of empirical laws would never have been achieved without the unifying aid of a molecular doctrine at the center; note the recent remarks on scope. But we might still delete the molecular doctrine once it has thus served its heuristic purpose.

This reflection strengthens a natural suspicion: that the benefits conferred by the molecular doctrine give the physicist good reason to prize it, but afford no evidence of its truth. Though the doctrine succeed to perfection in its indirect bearing on observable reality, the question of its truth has to do rather with its direct claim on unobservable reality. Might the molecular doctrine not be ever so useful in organizing and extending our knowledge of the behavior of observable things, and yet be factually false?

One may question, on closer consideration, whether this is really an intelligible possibility. Let us reflect upon our words and how we learned them.

II. POSITS AND ANALOGIES

Words are human artifacts, meaningless save as our associating them with experience endows them with meaning. The word 'swarm' is initially meaningful to us through association with such experiences as that of a hovering swarm of gnats, or a swarm of dust motes in a shaft of sunlight. When we extend the word to desks and the like, we are engaged in drawing an analogy between swarms ordinarily so-called, on the one hand, and desks, and so forth, on the other. The word 'molecule' is then given meaning derivatively: having conceived of desks analogically as swarms, we imagine molecules as the things the desks are swarms of.

The purported question of fact, the question whether the familiar objects around us are really swarms of subvisible particles in vibration, now begins to waver and dissolve. If the words involved here make sense only by analogy, then the only question of fact is the question how good an analogy there is between the behavior of a desk or the like and the behavior, for example, of a swarm of gnats. What had seemed a direct bearing of the molecular doctrine upon reality has now dwindled to an analogy.

Even this analogical content, moreover, is incidental, variable, and at length dispensable. In particular the analogy between the swarming of the molecules of a solid and the swarming of gnats is only moderately faithful; a supplementary aid to appreciating the dynamics of the molecules of a solid is found in the analogy of a stack of bedsprings. In another and more recondite part of physics, the theory of light, the tenuousness of analogy is notorious: the analogy of particles is useful up to a point and the analogy of waves is useful up to a point, but neither suffices to the exclusion of the other. Faithful analogies are an aid to the physicist's early progress in an unaccustomed medium, but, like water-wings, they are an aid which he learns to get along without.

In section I we contrasted a direct and an indirect bearing of the molecular doctrine upon reality. But the direct bearing has not withstood scrutiny. Where there had at first seemed to be an undecidable question of unobservable fact, we now find mere analogy at most and not necessarily that. So the only way in which we now find the molecular doctrine genuinely to bear upon reality is the indirect way, via implications in observable phenomena.

The effect of this conclusion upon the status of molecules is that they lose even the dignity of inferred or hypothetical entities which may or may not really be there. The very sentences which seem to propound them and treat of them are gibberish by themselves, and indirectly significant only as contributary clauses of an inclusive system which does also treat of the real. The molecular physicist is, like all of us, concerned with commonplace reality, and merely finds that he can simplify his laws by positing an esoteric supplement to the exoteric universe. He can devise simpler laws for this enriched universe, this "sesquiverse" of his own decree, than he has been able to devise for its real or original portion alone.

In section I we imagined deleting the molecular doctrine from the midst of the derivative body of physical theory. From our present vantage point, however, we see that operation as insignificant; there is no substantive doctrine of molecules to delete. The sentences which seem to propound molecules are just devices for organizing the significant sentences of physical theory. No matter if physics makes molecules or other insensible particles seem more fundamental than the objects of common sense; the particles are posited for the sake of a simple physics.

The tendency of our own reflections has been, conversely, to belittle molecules and their ilk, leaving common-sense bodies supreme. Still, it may now be protested, this invidious contrast is unwarranted. What are given in sensation are variformed and varicolored visual patches, varitextured and varitemperatured tactual feels, and an assortment of tones, tastes, smells, and other odds and ends; desks are no more to be found among these data than molecules. If we have evidence for the existence of the bodies of common sense, we have it only in the way in which we may be said to have evidence for the existence of molecules. The positing of either sort of body is good science insofar merely as it helps us formulate our laws—laws whose ultimate evidence lies in the sense data of the past, and whose ultimate vindication lies in anticipation of sense data of the future. The positing of molecules differs from the positing of the bodies of common sense mainly in degree of sophistication. In whatever sense the molecules in my desk are unreal and a figment of the imagination of the scientist, in that sense the desk itself is unreal and a figment of the imagination of the race.

This double verdict of unreality leaves us nothing, evidently, but the raw sense data themselves. It leaves each of us, indeed, nothing but his own sense data; for the assumption of there being other persons has no better support than has the assumption of there being any other sorts of external objects. It leaves each of us in the position of solipsism, according to which there is nobody else in the world, nor indeed any world but the pageant of one's own sense data.

III. RESTITUTION

Surely now we have been caught up in a wrong line of reasoning. Not only is the conclusion bizarre; it vitiates the very considerations that lead to it. We cannot properly represent man as inventing a myth of physical objects to fit past and present sense data, for past ones are lost except to memory; and memory, far from being a straightforward register of past sense data, usually depends on past posits of physical objects. The positing of physical objects must be seen not as an *ex post facto* systematization of data, but as a move prior to which no appreciable data would be available to systematize.

Something went wrong with our standard of reality. We became doubtful of the reality of molecules because the physicist's statement that there are molecules took on the aspect of a mere technical convenience in smoothing the laws of physics. Next we noted that common-sense bodies are epistemologically much on a par with the molecules, and inferred the unreality of the common-sense bodies themselves. Here our bemusement becomes visible. Unless we change meanings in midstream, the familiar bodies around us are as real as can be; and it smacks of a contradiction in terms to conclude otherwise. Having noted that man has no evidence for the existence of bodies beyond the fact that their assumption helps him organize experience, we should have done well, instead of disclaiming evidence for the existence of bodies, to conclude: such, then, at bottom, is what evidence is, both for ordinary bodies and for molecules.

This point about evidence does not upset the evidential priority of sense data. On the contrary, the point about evidence is precisely that the testimony of the senses *does* (contrary to Berkeley's notion) count as evidence for bodies, such being (as Samuel Johnson perceived) just the sort of thing that evidence is. We can continue to recognize, as in section II, that molecules and even the gross bodies of common sense are simply posited in the course of organizing our responses to stimulation; but a moral to draw from our reconsideration of the terms 'reality' and 'evidence' is that posits are not *ipso facto* unreal. The benefits of the molecular doctrine which so impressed us in section I, and the manifest benefits of the aboriginal posit of ordinary bodies, are the best evidence of reality we can ask (pending, of course, evidence of the same sort for some alternative ontology).

Sense data are posits too. They are posits of psychological theory, but not, on that account, unreal. The sense datum may be construed as a hypothetical component of subjective experience standing in closest possible correspondence to the experimentally measurable conditions of physical stimulation of the end organs. In seeking to isolate sense data we engage in empirical psychology, associating physical stimuli with human resources. I shall not guess how useful the positing of sense data may be for psychological theory, or more specifically for a psychologically grounded theory of evidence, nor what detailed traits may profitably be postulated concerning them. In our flight from the fictitious to the real, in any event, we have come full circle.

Sense data, if they are to be posited at all, are fundamental in one respect; the small particles of physics are fundamental in a second respect, and common-sense bodies in a third. Sense data are *evidentially* fundamental: every man is beholden to his senses for every hint of bodies. The physical particles are *naturally* fundamental, in this kind of way: laws of behavior of those particles afford, so far as we know, the simplest formulation of a general theory of what happens. Common-sense bodies, finally, are *conceptually* fundamental: it is by reference to them that the very notions of reality and evidence are acquired, and that the concepts which have to do with physical particles or even with sense data tend to be framed and phrased. But these three types of priority must not be viewed as somehow determining three competing, self-sufficient conceptual schemes. Our one serious conceptual scheme is the inclusive, evolving one of science, which we inherit and, in our several small ways, help to improve.

IV. WORKING FROM WITHIN

It is by thinking within this unitary conceptual scheme itself, thinking about the processes of the physical world, that we come to appreciate that the world can be evidenced only through stimulation of our senses. It is by thinking within the same conceptual scheme that we come to appreciate that language, being a social art, is learned primarily with reference to intersubjectively conspicuous objects, and hence that such objects are bound to be central conceptually. Both of these *aperçus* are part of the scientific understanding of the scientific enterprise; not prior to it. Insofar as they help the scientist to proceed more knowingly about his business, science is using its findings to improve its own techniques. Epistemology, on this view, is not logically prior somehow to common sense or to the refined common sense which is science; it is part rather of the overall scientific enterprise, an enterprise which Neurath has likened to that of rebuilding a ship while staying afloat in it.

Epistemology, so conceived, continues to probe the sensory evidence for discourse about the world; but it no longer seeks to relate such discourse somehow to an imaginary and impossible sense-datum language. Rather it faces the fact that society teaches us our physicalistic language by training us to associate various physicalistic sentences directly, in multifarious ways, with irritations of our sensory surfaces, and by training us also to associate various such sentences with one another.

The complex totality of such associations is a fluctuating field of force. Some sentences about bodies are, for one person or for many, firmly conditioned one by one to sensory stimulation of specifiable sorts. Roughly specifiable sequences of nerve hits can confirm us in statements about having had breakfast, or there being a brick house on Elm Street, beyond the power of secondary associations with other sentences to add or detract. But there is in this respect a grading-off from one example to another. Many sentences even about common-sense bodies rest wholly on indirect evidence; witness the statement that one of the pennies now in my pocket was in my pocket last week. Conversely, sentences even about electrons are sometimes directly conditioned to sensory stimulation, for example, via the cloud chamber. The status of a given sentence, in point of direct or indirect connection with the senses, can change as one's experience accumulates; thus a man's first confrontation with a cloud chamber may forge a direct sensory link to some sentences which hitherto bore, for him, only the most indirect sensory relevances. Moreover the sensory relevance of sentences will differ widely from person to person; uniformity comes only where the pressure for communication comes.

Statements about bodies, common-sense or recondite, thus commonly make little or no empirical sense except as bits of a collectively significant containing system. Various statements can surely be supplanted by their negations, without conflict with any possible sensory contingency, provided that we revise other portions of our science in compensatory ways. Science is empirically underdetermined: there is slack. What can be said about the hypothetical particles of physics is underdetermined by what can be said about sensible bodies, and what can be said about these is underdetermined by the stimulation of our surfaces. An inkling of this circumstance has doubtless fostered the tendency to look upon the hypothetical particles of physics as more of a fiction than sensible bodies, and these as more of a fiction than sense data. But the tendency is a perverse one, for it ascribes full reality only to a domain of objects for which there is no autonomous system of discourse at all.

Better simply to explore, realistically, the less-than-rigid connections that obtain between sensory stimulus and physical doctrine, without viewing this want of rigidity as impugning the physical doctrine. Benefits of the sort recounted in section I are what count for the molecular doctrine or any, and we can hope for no surer touchstone of reality. We can hope to improve our physics by seeking the same sorts of benefits in fuller measure, and we may even facilitate such endeavors by better understanding the degrees of freedom that prevail between stimulatory evidence and physical doctrine. But as a medium for such epistemological inquiry we can choose no better than the self-same world theory which we are trying to improve, this being the best available at the time.

On Meaning-Dependence

5. A DOCTRINE ABOUT MEANING-DEPENDENCE

The doctrine I will now consider has emerged in recent years as a major alternative to Logical Positivism. Its proponents are influenced by a study of the history of science. They object to the abstractness of the Positivist account, to the intellectual strait jacket they say it imposes on science by the use of symbolic logic, and to what they regard as its lack of concern with actual scientific procedures. In particular, they emphasize, attention to the history of science will show that the terms employed are "burdened" or "laden" with the scientific theory in which they are used. Accordingly, to understand what such terms mean, one must learn that theory; and the meaning of the terms will change if the theory is modified or replaced by another. To say this is to reject a fundamental assumption of Positivism—namely, that there are, and indeed must be, terms in a theory (the nontheoretical terms) whose meanings can be given independently of the theory and can remain constant from theory to theory. The doctrine I want to consider, then, contains the following theses:

1) The meaning of a term that occurs in a scientific theory is dependent upon principles of that theory, so that to know what the term means requires a knowledge of that theory.

2) The meaning of a term that occurs in a scientific theory changes when the theory is modified or replaced by another theory in which that term also occurs.

Many have been willing to defend some form of this view for at least some terms. I shall consider this doctrine in a form held by

Excerpted and reprinted by kind permission of the author and publishers from *Concepts of Science* by Peter Achinstein. Baltimore: The Johns Hopkins Press, 1968, pp. 91–105. Copyright by The Johns Hopkins Press, 1968.

Feyerabend.[1] In this form the doctrine is supposed to apply to *all* terms utilized in a theory, including all those used for describing observations which support the theory. The meanings of the terms in the theory are supposed to be entirely dependent upon the theory; there are no theory-neutral or partly theory-neutral terms in a theory. Indeed, Feyerabend claims, "the description of every single fact [is] dependent on *some* theory. . . . The meaning of every term we use depends upon the theoretical context in which it occurs. Words do not 'mean' something in isolation; they attain their meanings by being part of a theoretical system." [2] Moreover, when a new theory emerges to replace the old one, the terms involved will change in such a way that there will be an "elimination of the old meanings," and the same term, although employed in both cases, will express two different and "incommensurable" concepts. "Introducing a new theory," Feyerabend writes, "involves changes of outlook both with respect to the observable and with respect to the unobservable features of the world, and corresponding changes in the meanings of even the most 'fundamental' terms of the language employed." [3]

There are, I think, important insights here, but, stated in the manner proposed by Feyerabend, the doctrine has several unacceptable consequences. First, if it were true, no theory could contradict another. Consider the Bohr theory of the atom, which assumes that electrons revolve about the nucleus of an atom in such a way that their orbital angular momentum is quantized (it is a whole multiple of $h/2\pi$, where h is Planck's constant); it also assumes that energy is radiated or absorbed by the atom only when an electron jumps from one stable orbit to another and that this energy is also quantized. When the Bohr theory claims that angular momentum and radiant energy of electrons cannot have continuous values but must be quantized, it *denies* the assumption of classical electrodynamics that angular momentum and radiant energy of electrons can have continuous values. Bohr himself writes as follows:

> Now the essential point in Planck's theory of radiation is that the energy radiation from an atomic system does not take place in the continuous way assumed in the ordinary electrodynamics, but that it, on the contrary, takes place in distinctly separated emissions. . . .[4]

[1] P. K. Feyerabend, "Explanation, Reduction, and Empiricism," *Minnesota Studies in the Philosophy of Science,* ed. H. Feigl and G. Maxwell (Minneapolis, 1962), III, 28–97. See also "Problems of Empiricism," *Beyond the Edge of Certainty,* ed. R. G. Colodny (Englewood Cliffs, N.J.: Prentice-Hall, Inc., 1965), pp. 145–260. Cf. Thomas S. Kuhn, *The Structure of Scientific Revolutions* (Chicago, 1962).

[2] "Problems of Empiricism," pp. 175, 180.

[3] "Explanation, Reduction, and Empiricism," p. 29.

[4] Niels Bohr, "On the Constitution of Atoms and Molecules," *Philosophical Magazine,* 26 (1913), 4.

All such denials are mere illusions, on the present account, for the terms in the Bohr theory have different meanings from those in classical electrodynamics. They express "incommensurable" concepts. When Bohr asserts that "the energy radiation from an atomic system does not take place in [a] continuous way" and when a classical theorist asserts that "energy radiation from an atomic system takes place in a continuous way," they are using words with different meanings and so cannot be contradicting each other. Nor could other terms from the two theories be used to express their disagreement, since all pairs of terms from different theories express "incommensurable" concepts. This is an absurd consequence.

It eliminates or at least seriously weakens the concept of negation. When does the meaning of the terms used in asserting something become dependent upon what is asserted? Perhaps this is so for any assertion. If I say that an atomic system radiates energy continuously and you say that an atomic system does not radiate energy continuously, then by an atomic system I mean (among other things) something which radiates energy continuously and you mean something else. According to this approach, if I assert p and you assert *not-p*, we are not and cannot be disagreeing, because the terms in my assertion are *p-laden* and so mean one thing, whereas those in *not-p* are *not-p-laden* and so mean another. *Not-p*, then, is not the negation of p. In short, negation is impossible! On the other hand, if every assertion does not burden its terms with what is asserted, then which ones do and why? The answers are not forthcoming.

Feyerabend does claim that disagreement between two theories can be established without appealing to the meanings of terms and hence without assuming identity in meanings. His proposal, although vaguely formulated, seems to be this: two theories are incompatible or in basic disagreement if (and possibly only if) a lack of isomorphism exists between them. Feyerabend seems to be speaking of isomorphism between the elements described by the theory, but he may also have in mind isomorphism with respect to the theoretical sentences themselves; and, he insists, lack of isomorphism can be established without appeal to meanings.[5]

Lack of isomorphism in either case is not a *sufficient* condition for what would normally be regarded as a disagreement between theories. Keynesian theory and quantum theory are lacking in isomorphism with respect to the elements they describe and also with respect to the formal properties of the sentences used in describing them. They are not, however, in disagreement, or agreement either. (They are genuinely incommensurable.) Nor is lack of isomorphism a *necessary* condition for what would normally be regarded as a disagreement between theories. This can be seen in a simple way if we consider the following sets of propositions that, by normal standards, would be considered incompatible:

Theory 1	Theory 2
All planets in our solar system attract each other	All planets in our solar system repel each other
There exist nine planets in our solar system	There exist nine planets in our solar system
No planets that attract each other repel each other	No planets that attract each other repel each other

These sets of propositions, although incompatible, are formally speaking isomorphic, and there is an isomorphism between the postulated elements (the planets and the relationship of attracting and repelling).

In reply to an earlier criticism of mine,[6] Feyerabend claims that in certain cases two theories may disagree and yet have terms with common meanings; but these he seems to regard as trivial and uninteresting cases, for example, classical mechanics and a theory imagined to be just like it except for a slight change in the strength of the gravitational potential.[7] He then explains how differences in meaning can be ascertained. A theory T will provide "rules according to which objects or events are collected into classes. We may say that such rules determine concepts or kinds of objects." [8] If a new theory T' simply produces changes within the extension of these classes, as happens presumably in celestial mechanics when a slight change is made in the strength of the gravitational potential, there is no change in meaning. However, if the new theory "entails that all the concepts of the preceding theory have extension zero or if it introduces rules which cannot be interpreted as attributing specific properties to objects within already existing classes, but which change the system of classes itself," then terms in the two theories have different meanings.[9]

I find this proposal unworkable, for how can the new theory entail that the concepts of the preceding theory have extension zero unless there are terms in the two theories with common meanings? If we rely on the second part of the alternation to ascertain differences in meaning, difficulties also emerge. As Dudley Shapere has pointed out in discussing Feyerabend's present proposal, given the principles of a theory, in general there will be a number of ways we might collect the items described by these principles into classes, depending on our pur-

[6] Peter Achinstein, "On the Meaning of Scientific Terms," *Journal of Philosophy*, 61 (1964), 497–509.

[7] Feyerabend, "On the 'Meaning' of Scientific Terms," *Journal of Philosophy*, 62 (1965), 266–74.

[8] *Ibid.*, p. 268.

[9] *Ibid.*

poses.[10] If we make our class-descriptions sufficiently general (for example, "physical processes," "physical objects"), then any two theories, no matter how much in disagreement, will involve the same "system of classes" and will simply be attributing different "specific properties to objects within already existing classes." If we make our class-descriptions sufficiently specific, then two theories that disagree, even in trivial ways, would involve different systems of classes. For example, celestial mechanics and a theory indistinguishable from it, except for a difference in the strength of the gravitational potential, can be construed as generating different classes (different "kinds of objects"): the former postulates a class of bodies subject to one force law; the latter, a class of bodies subject to a different one. In short, Feyerabend's present suggestion does not yield the sort of method he desires for determining meaning-changes; nor does it even allow him to distinguish trivial cases where meanings are constant from interesting ones where they are not.

The second unacceptable consequence of Feyerabend's general doctrine is related to the first and can be stated briefly. Not only is any disagreement impossible for proponents of two different theories but for the same reason, so is any agreement. It will be impossible for theorists to agree even on a description of the data to be explained by their respective theories, for in such a description all the terms employed will depend for their meanings upon the given theory. But if there can be no agreement and no disagreement either, in what sense can two different theories be about the same thing? In what sense can two different theories be (as Feyerabend calls them) *alternatives?*

There is a third untoward consequence of this doctrine. When it is claimed that the meaning of a term in a theory depends entirely on principles of that theory, the suggestion is that these principles in effect say what the term means. If so, then such principles would always be construable as analytic, that is, as defensible solely by appeal to the meanings of their constituent terms. The claim of the Bohr theory that the electron's angular momentum is quantized could be defended by appeal solely to the fact that "electron" and "angular momentum" mean just those sorts of things such that the angular momentum of the electron is quantized.

To this it might be replied that although most of the principles of the theory will turn out to be analytic its existence claims will not. For example, the claim that there exist atoms satisfying the principles of the Bohr theory will be empirical. On this construal, the postulates of the theory are analytic, and the question whether anything satisfies these postulates is empirical.

One might object to this extreme manner of reconstructing a theory

[10] Dudley Shapere, "Meaning and Scientific Change," *Mind and Cosmos*, ed. R. Colodny (Pittsburgh, 1966), pp. 41–85.

so that many sentences one would normally take to be expressing only empirical propositions, sentences scientists treat as such, now express statements defensible simply by appeal to the meanings of their terms. But there is a more serious problem. If all terms employed by the theory are theory-laden, including the terms employed in describing observations made in connection with the theory, then a description of any such observations will presuppose the theory. No observed item could be described as failing to satisfy the postulates of the theory. Nor could any other theory T' be used to show that items exist that do not satisfy the theory T; since the concepts of any such T' are incommensurable with those of T, T' cannot be used to show that items exist which satisfy the negation of T. This would make the existence claims of a theory irrefutable by any observations that could be described or by other theories based on such observations. In what sense, then, could such claims be called empirical? Even independently of language, observations could not be made that would refute existence claims of a theory, since, for Feyerabend, observations themselves must always presuppose a theory; they are theory-dependent. If they are made presupposing T, then T could not be refuted; nor could it be refuted if T' were presupposed, since T' and T cannot be conflicting theories.

The fourth consequence I find unacceptable is that if the doctrine were true, a person could not learn a theory by having it explained to him using any words whose meanings he understands before he learns the theory. Consider the terms "electron," "electron orbit," "angular momentum," and "radiant energy," which appear in the Bohr theory. According to thesis 1 (page 162), in order to know what these terms mean, I must know what the Bohr theory asserts. One of the central principles in this theory can be expressed like this:

> Of all the electron orbits only those orbits are permissible for which the angular momentum of the electron is a whole multiple of $h/2\pi$, and there is no radiant energy while the electron remains in any one of these permissible orbits.

But how can I know what is asserted here unless I know what the terms "electron," "electron orbit," "angular momentum," and "radiant energy" mean? It is useless to appeal to what they mean *in other theories;* for, on this doctrine, what they mean in two theories is different and "incommensurable." Nor can I explain what they mean by using words whose meanings are independent of any theory, since there are supposed to be no theory-neutral terms. Therefore, I cannot use any words whose meanings I understand in order to learn the meanings of the terms in the theory. The only thing I can do is try to learn the meanings *extralinguistically.* I must watch what those who use the theory do in their laboratories, the sorts of items to which they apply their terms, and so forth. I must learn each new theory like a child first learning language (rather than like someone learning more of his own language or a second language after learning a first one). Perhaps it would be possible (though, I suspect, exceedingly difficult) to learn scientific theories this way. What I find unacceptable is the consequence that they *must* be learned this way. In actual practice at least some if not most terms in a new theory are explained to those learning the theory by using words whose meanings the learners already know. Those who learn the Bohr theory do not do so only by observing the behavior of those who already know it and the phenomena such people may point out.

6. CRITERIA FOR MEANING-DEPENDENCE AND FOR CHANGE IN MEANING

The four consequences I have noted all stem from the same source, dubious assumptions about *meaning*. Feyerabend fails to propose any theory of meaning, or at least any adequate one, that will support his claim that the meaning of a term in a theory depends upon principles of that theory and will change when the theory is altered or replaced. He does, of course, cite examples in propounding this thesis.[11] Some of these are convincing on intuitive grounds, but others are not. Still further examples that could be cited seem, intuitively, to violate this thesis.

On intuitive grounds, at least, it seems absurd to claim that one cannot understand what meaning is associated with terms such as "angular momentum," "radiant energy," and "electron" unless one knows the Bohr theory, since all these terms were in use prior to this theory and Bohr simply appropriated them for his theory. On intuitive grounds, it also seems absurd to say that the terms "angular momentum" and "radiant energy" suffered a complete change in meaning when Bohr postulated that the quantities they designate cannot have continuous values but must be quantized. I shall say something about these intuitive grounds and afterward show how the position developed in the chapters on definition can be used to explain and defend these intuitions.

One set of considerations relevant for deciding whether a term means the same in two theories and also whether its meaning depends upon a theory are "behavioral." Suppose a term used in a well-known theory T comes to be used also in a different theory T'. And suppose T' is presented in such a way that an explanation of the semantical

[11] See "Explanation, Reduction, and Empiricism," and "On the 'Meaning' of Scientific Terms."

aspect of the use of this term is not given, nor are contexts supplied that would indicate very much if anything about this. Do those studying such a presentation of T', knowing the theory T, request an explanation for the meaning of this term, and if not, do they seem to misunderstand what T' is supposed to assert? Are there the usual signs of a lack of communication and understanding when explanations or appropriate contexts are not supplied? If not, this provides at least some ground for saying that the meaning of the term can be understood without learning T' and that it means the same in both theories.

These considerations lend weight to the claim that the meaning of a term such as "angular momentum" could be learned independently of the Bohr theory and is the same in that theory as in classical mechanics. Bohr, who used this term, felt no need to define or redefine it or to present contexts that would make the semantical aspect of its use apparent; nor do texts on atomic physics, which propound and explain this theory for students already acquainted with classical mechanics.

Another consideration concerns the manner in which claims involving the term are held to be settleable. Suppose that one theory appears to be asserting, and another to be denying, that X's have P. If the terms "X" and "P" have the same meaning in each theory, then there should be facts that are or could be relevant in settling whether X's do have P, as "X" and "P" are used in each theory. If proponents of these two theories cannot agree in principle on how to settle, or on what might be relevant for settling, whether X's have P, this suggests that they are using the term "X" or the term "P" (or both) in different ways.

There are a number of facts that were taken to provide at least some support for the Bohr theory against that of classical electrodynamics. One is that spectra of the elements are not continuous but discontinuous: if, for example, an element (such as hydrogen) is suitably excited by having a current passed through it, it gives off light which, when analyzed with a spectroscope, is found to consist of a series of discrete lines of certain wavelengths. The fact that Bohr could deduce from this theory the exact wavelengths of these lines, whereas classical electrodynamics could not even explain their existence, was taken as some support for Bohr's assumptions regarding angular momentum and radiant energy against those of the classical theory. Another fact concerns resonance potential experiments. Through a tube containing hydrogen at low pressure a current is passed from a filament to a plate. As the voltage is increased, the current is increased to a certain point after which it drops off (where this is repeated for higher voltages): The drop in current is taken as indicating that the energy of electrons is just the right amount (the "quantum" or multiple of it) to raise the hydrogen atoms from a normal state to an excited state. This was also claimed to provide some support for Bohr's assumption (against that of classical theory) regarding quantization of energy within the atom.

A third consideration for deciding whether a term means the same in two theories and whether its meaning depends on a certain theory is suggested by Putnam.[12] Suppose I know or am told how the term "X" is now used in a set of circumstances C (or in connection with a set of assumptions C). That is, suppose I know the sorts of items, actual and hypothetical, to which the term is applicable and also in virtue of what it is applicable. If, from this knowledge alone, it is possible for me to predict how the term "X" would be used in circumstances C' (or in connection with a different set of assumptions C'), this counts against speaking of a different meaning in these circumstances. To the degree that its use in circumstances C' is unpredictable from its use in present circumstances C, we can speak of a different meaning. For example, if in the future our scientific theories remain the same but the term "solid" comes to be used in such a way that items in gaseous form, and only these, are regarded as standard examples of solids, this would be a radically different meaning, since this use of "solid" would be wholly unpredictable on the basis of the present one. Furthermore, if, on the basis of the present use of "X" in principles constituting a theory T (that is, if knowing the sorts of items to which it is applicable and in virtue of what it is applicable), I could predict how it would be used in principles constituting a theory T', then this provides some grounds for saying that the meaning of the term "X" can be known independently of T'. Of course to know that "X" is being used in the same way in T' and T, I may have to learn T'. But it does not follow that to know what "X" means, I will have to learn T' (or T for that matter).

These considerations also suggest that the meanings of certain terms in the Bohr theory are the same as in classical electrodynamics and can be learned independently of the Bohr theory. We begin with the term "angular momentum" used in a theory that assumed that all atomic states are subject to principles of classical mechanics and that angular momentum in the case of electrons can have continuous values. From this it was certainly predictable how the term would be used (that is, how it would be defined, how its values could be determined) in a theory asserting that only stationary states of the atom are subject to principles of classical mechanics and that the angular momentum of electrons can have only discrete values proportional to $h/2\pi$. One could reasonably have supposed, for example, that the angular momentum of the electron in the latter case (assuming a circular orbit of radius r) would be given by the expression mvr, where m is its mass and v its velocity.

All three considerations mentioned allow for the possibility that terms in two theories can have the same meaning and that the meaning

[12] "Dreaming and 'Depth Grammar,'" p. 223.

of a term used in a theory can be known without knowing the theory. Now I want to defend this by appeal to the position developed in the chapters on definition. According to this position, some properties attributed to an item X (or some conditions satisfied by X) are logically necessary or sufficient for X; others may be relevant. Among the latter I distinguished the semantically relevant from the nonsemantically relevant, while recognizing that some are not happily classifiable in either way. Properties semantically relevant for X (which include those that are logically necessary or sufficient) have a particularly intimate connection with the meaning of the term "X" (with the semantical aspect of use) as other properties do not.*

Consider a term "X" and the properties or conditions semantically relevant for X. It is perfectly possible that there be two different theories in which the term "X" is used, where the same set of semantically relevant properties of X (or conditions for X) are presupposed in each theory (even though other properties attributed to X by these theories, properties not semantically relevant for X, might be different). If so, the term "X" would not mean something different in each theory.

Suppose $P_1, \ldots, P_n$ represent all the semantically relevant properties of X. Suppose that T is a theory that says that X's have $Q_1, \ldots, Q_m$, where each of these is semantically independent of (has no semantical relevance for) $P_1, \ldots, P_n$. Suppose T' is a theory that says that X's have $Q'_1, \ldots, Q'_m$, where each of these is semantically independent of each of $P_1 \ldots, P_n$, but where some or all of the Q''s are logically incompatible with the corresponding Q's. If so, theories T and T' would be incompatible. But if $P_1, \ldots, P_n$ represent all the semantically relevant properties of X, as "X" is used in both theories, then "X" would mean the same in T' as in T, and its meaning could be learned independently of T' (and of T). There is no *a priori* reason why this cannot happen. Although this is an oversimplified schematization, I would claim that the semantically relevant properties of certain items designated by terms in theories can be known without learning those theories and are the same in other theories in which the terms appear.

This is so for "position" and "time" in the Bohr theory and in classical mechanics. Both theories use the same set of spatial and temporal concepts. Whatever conditions are semantically relevant for a particle having a position at a given time are the same for the Bohr theory as for classical mechanics. And the definitions of the terms "velocity" as "time rate of change of position" and "acceleration" as "time rate of change of velocity" provide logically necessary and sufficient conditions

---

* Roughly, a predicate F is semantically relevant to the predicate G if the knowledge that an object has F is important in determining whether it has G. For a complete presentation of the ideas involved, see Chapter 1 of Achinstein's *Concepts of Science*. —Ed.

each of which is exactly the same in the two theories. Indeed, if the concepts I mention were not the same in both theories, Bohr could not have used principles of classical mechanics with these concepts in order to derive his expression for the total energy of the electron in the hydrogen atom. One of the things Bohr did was use Newton's second law of motion, $F = ma = mv^2/r$, where F is the force of attraction on the electron by the nucleus, a is its centripetal acceleration, and v its velocity.[13] From this, together with Coulomb's law of attraction, the kinetic energy formula in classical mechanics, and Bohr's assumptions regarding quantization of angular momentum and energy, it was possible to arrive at a formula relating the total energy of the electron to its mass, charge, the number of its orbit, and Planck's constant.

Not only can two theories use a term with the same set of semantically relevant properties or conditions, but, as with terms such as "velocity," "acceleration," and "angular momentum" in the Bohr theory, it is also possible for one to learn what these properties or conditions are without learning the theory. In short, some terms employed by a theory T will be such that none of the semantically relevant properties attributed to the items they designate will be attributed to them by theory T itself or only by T (but perhaps by other theories or even independently of any theory). Moreover, some terms may be employed in two theories in such a way that the set of semantically relevant properties associated with items they designate are the same in each theory. In the first case, it is proper to speak of learning what the term means (learning the semantical aspect of its use) independently of the theory in question; in the second, it is proper to speak of the term as meaning the same in both theories (in both theories the semantical aspect of use is the same). It is proper to speak in these ways even though a knowledge of X's properties that are not semantically relevant may require learning a theory or theories, and even though two theories may attribute to X different (and even conflicting) properties that are not semantically relevant.

Let me relate this to the three intuitive considerations governing meaning-dependence and meaning-change noted earlier. These concerned behavior, settleability, and predictability. The reason why such considerations are relevant can be understood on the basis of the view I have been expounding. It is because of knowledge (or lack of it) of semantically relevant properties of X that those who learn the new theory exhibit none (or all or some) of the "behavioral" symptoms of understanding when meeting the term in a new theory. Moreover, if proponents of two theories cannot agree in principle on how to settle, or on what might be relevant in settling, whether X's have P, this

[13] Niels Bohr, "On the Constitution of Atoms and Molecules," p. 478.

suggests that they disagree on that in virtue of which something is classifiable as X, P, or both. In short, this suggests that they cannot agree on the semantically relevant properties of X, the semantically relevant conditions for P, or both. This is why lack of agreement on what might settle whether X is P is relevant to the question of whether the terms mean the same for both parties. Finally, one's prediction about the use of a term in a new context on the basis of its use in the present one derives from a knowledge of what properties or conditions are semantically relevant in the latter and whether the same or similar ones are apt to be semantically relevant in the new context as well.

I have argued that it is possible for there to be terms in a theory whose meanings can be learned without learning the theory and can be the same as those of terms in a different theory. But two other sorts of cases should be noted. There may be terms in a theory T that designate items some of whose semantically relevant properties will be attributed to them by T and only by T, whereas others will be attributed to them independently of T, perhaps by some theory T'. For such terms, neither of Feyerabend's labels "meaning-dependent (upon T)" or "meaning-independent (of T)" may be appropriate (where these are construed in his sense, requiring complete dependence or independence). It may be best to avoid these oversimplified classifications and indicate that certain semantical aspects of the use of these terms depend upon T, whereas others do not. Similarly, there may be terms used in two theories T and T' in such a way that although some of the semantically relevant properties attributed to the items they designate are the same in each theory, others are not. For such items, neither the label "same meaning" nor "different meaning" may be appropriate (where, again, these are understood as requiring complete sameness or difference).

The second type of case to be noted does fit Feyerabend's account: some terms utilized by a theory will be such that all or most of the properties or conditions semantically relevant for X will be attributed to X by that particular theory and only by that theory. Even though the same term may be used in another theory, the properties semantically relevant for X will be substantially different in each theory. Consider the term "entropy" in classical thermodynamics. Many if not most of the conditions satisfied by entropy and likely to be treated as semantically relevant would be conditions specified by, or consequences of, the principles of classical thermodynamics; for example, that the entropy of an isolated system increases in any process or remains constant in equilibrium, that entropy measures the loss of available energy, and so forth. Now statistical mechanics, which also uses the term "entropy," goes beyond the assumptions of classical thermodynamics by introducing postulates concerning the molecular microstructure of a thermodynamic system. A complete specification of the position and velocity (within certain limits) of each particular molecule defines a *microstate* of the system; and the specification of the number of molecules with each given set of dynamical properties defines a *macrostate*. According to the Boltzmann postulate, the entropy of a thermodynamic system (as defined in classical thermodynamics) is proportional to the logarithm of the number of microstates belonging to the given macrostate of the system. This postulate is expressed symbolically by $S = k \log W$. One way to construe this (adopted by Boltzmann) is to take S as classical entropy. If so, then statistical mechanics attributes certain new properties to entropy, while retaining many others from classical thermodynamics. However, another procedure (discussed by Fowler) is to treat $S = k \log W$ as providing a logically necessary and sufficient condition for entropy.[14] If this is adopted, then, I think, a "different meaning" label would be warranted, since the present condition is not logically necessary, logically sufficient, or otherwise semantically relevant for classical entropy.

Therefore, cases exist in which Feyerabend's labels "meaning-dependent (upon T)" and "different meaning (in T and T')" are appropriate. The merit of his position lies in reminding us that there are such cases, that terms *can* depend completely for their meanings upon a theory and suffer extensive meaning changes when theories are modified or replaced. However, there are also cases in which the labels "meaning-independent (of T)" and "same meaning (in T and T')" are warranted, and some where none of these labels fits. The view under discussion fails to recognize the latter sorts of cases. It fails in this because it treats *all* properties attributed to X by a theory as if they were semantically relevant for X.

The view also assumes that any term used by the proponent of a theory to describe *observations* made in connection with that theory will be a term whose semantically relevant conditions are supplied by the theory. In short, it assumes that any such term will depend completely for its meaning upon that theory. This position lies at the opposite pole from the Positivist view discussed earlier. According to the latter, to describe what is observed, a special "observational" vocabulary is required whose terms in no way depend upon a theory and, indeed, can be common to all theories. The position that should be taken, I suggest, lies somewhere between these extremes. . . . There are numerous ways to describe one and the same set of observations. Some may employ terms that appear in the fundamental principles of a theory, others may not; but even if terms of the former sort are used, it does not necessarily follow that they will be entirely or even partially dependent upon the theory for their meaning.

In the previous section I noted four unacceptable consequences of

[14] R. H. Fowler, *Statistical Mechanics* (Cambridge, England, 1929).

Feyerabend's view. These were that if it were true, no theory could contradict another; any agreement between theories would be impossible; theories would be analytic; and it would be necessary to learn a theory completely extralinguistically. The position I have defended avoids these consequences. It is possible for two different theories to contradict each other since it is possible for them to be using terms with common meanings. For the same reason, some agreement between theories is possible, for example, agreement on at least some description of the data to be explained. Moreover, since not all properties attributed to X by a theory need be semantically relevant for X, it is not necessary that the principles of the theory be analytic. Finally, it is possible for a person to learn a theory by having it explained to him in words he understands before he learns the theory. For, at least in the case of some terms, since many, and perhaps all, of the semantically relevant properties of the items they designate may be attributed to them independently of the theory, it is possible to learn much about the semantical aspect of their use without presupposing the theory.

PAUL K. FEYERABEND

On the "Meaning" of Scientific Terms

I

In his criticism of Ryle, Hanson, and me, Peter Achinstein[1] notices "considerable oversimplification[s]" and discusses "paradoxical consequences" arising therefrom. . . . He points out that meanings do not always change with the theory to which they belong and suggests the existence of "various kinds and degrees of dependence as well as independence" between terms and theories. . . . He believes that awareness of these different kinds and degrees will eliminate the paradoxes and support two assumptions he finds plausible, namely "A. . . . it is possible to understand at least some terms employed in a . . . theory before (and hence without) learning the principles of that theory"; and "B. It must be possible for two theories employing many of the same terms to be incompatible . . . And this presupposes that at least some of the common terms have the same meaning in both theories" (499).

This belief of Achinstein's seems to be refuted by the existence of pairs of theories that may be regarded as competitors and yet do not share any element of meaning. Attention to "various kinds and degrees of dependence" clearly cannot eliminate such cases—it will rather bring them to the fore. Two examples exhibiting the property just described will be discussed in the next two sections. It will then be argued that, from the point of view of scientific progress, the examples are to be welcomed (IV). It will also be shown that B, despite its *prima facie* plausibility, is of a very dubious nature (V). Finally, we

Reprinted by kind permission of the author and publishers from the *Journal of Philosophy*, Vol. XII, No. 10 (May 13, 1965), 266–74.

[1] "On the Meaning of Scientific Terms," *Journal of Philosophy*, 61, 17 (Sept. 17, 1964): 497–509.

shall arrive at the result that, in the decision between competing theories, "meanings" play a negligible part and that attention to the "variety of kinds and degrees of dependence," while certainly populating the semantical zoo, does not solve a single philosophical problem.

II

The first example is the pair $[T, T']$, with T = classical celestial mechanics and T' = general theory of relativity. For the purpose of comparison I shall also consider $\bar{T} = T$ except for a slight change in the strength of the gravitation potential.

Now T and $\bar{T}$ are certainly different theories—in our universe, where no region is free from gravitational influence, no two predictions of $\bar{T}$ and T will coincide. Yet it would be rash to say that the transition $T \to \bar{T}$ involves a change of meaning. For though the *quantitative values* of the forces differ almost everywhere, there is no reason to assert that this is due to the action of different *kinds of entities*. After all, the existence of rubber bands of different strength does not indicate that there are various concepts of "rubber band." Nor does the existence of such a variety indicate that the notion of the surrounding space-time continuum is not well defined. There is nothing mysterious about such stability. The concept "rubber band"—or the more abstract concept "force"—*has been designed* to cover a great variety of entities, among them also entities of different strength. Hence, they are not affected by any transition leading from one element of their extension to another.

This example shows that a diagnosis of *stability of meaning* involves two elements. First, reference is made to rules according to which objects or events are collected into classes. We may say that such rules determine concepts or kinds of objects. Secondly, it is found that the changes brought about by a new point of view occur *within* the extension of these classes and, therefore, leave the concepts unchanged. Conversely, we shall diagnose a *change of meaning* either if a new theory entails that all concepts of the preceding theory have extension zero or if it introduces rules which cannot be interpreted as attributing specific properties to objects within already existing classes, but which change the system of classes itself.

It is important to realize that these two criteria lead to unambiguous results only if some further decisions are first made. Theories can be subjected to a variety of interpretations, and the relation of concepts to practice can also be seen in many different ways. Not every inter-

---

[2] Berkeley's notion of space as explained in *De Motu* was different from Newton's —but this difference appeared neither in experiment nor in the mathematical

pretation leaves its mark on current procedures. Interesting ideas may therefore be invisible to those who are concerned with the relation between existing formalisms and "experience" only.[2] It follows that we must (a) adopt a certain notion of "interpretation"; and (b) choose from among the various kinds of interpretations consistent with this notion the particular one we prefer. Questions concerning constancy or change of meaning have an unambiguous answer only *after* the decisions just described have been made. Otherwise we are dealing with pseudoproblems which, of course, we can "solve" or "refute" in any manner we please.

In what follows I shall decide (a) and (b) without giving detailed reasons for my decision. No such discussion is needed in the present brief note. All I intend to show now is that a position I hold can be presented coherently and that the alleged paradoxes it creates are harmless. I shall decide (a) by rejecting Platonism. This makes human practice the guide for conceptual considerations and the object of suggestions of conceptual reform. I shall decide (b) by adopting an epistemological realism. This means regarding theoretical principles as fundamental and giving secondary place to the "local grammar," that is, to those peculiarities of the usage of our terms which come forth in their application in concrete and, possibly, observable situations. It is intended to subject the local grammar to the theories we possess rather than to interpret the theories in the light of the knowledge—or alleged knowledge—that is expressed in our everyday actions. Or, to put it differently, we want to analyze, to explain, to justify, and perhaps occasionally to correct the "common knowledge" (which may also be the scientific knowledge of the preceding generation) by relating it to new theoretical ideas rather than to interpret such ideas as new ways of talking about what is already well known. This is the way in which fundamental revolutions have taken place in the seventeenth century, and again in the twentieth century; and this is also the way that a reasonable theory of knowledge invites us to take.[3]

Let us now apply these decisions to the case at hand. Will our diagnosis be one of change or of stability? And if it is a diagnosis of change, then what kind of change can we expect? We see at a glance that the spatiotemporal frameworks of T and T' certainly have little in common (three-dimensional Euclidian continuum with absolute time constituting a four-dimensional continuum not permitting any nonsingular metric in the case of T; four-dimensional Riemannian continuum

---

formalism accepted by either man. It consisted in an attitude influencing the *future development* of the theory of gravitation.

[3] See my "Problems of Empiricism," in *Beyond the Edge of Certainty*. Robert G. Colodny, ed. (Englewood Cliffs, N.J.: Prentice-Hall, Inc., 1965).

with nonsingular metric in the case of T': not even the over-all topology of the two spaces is the same). Stability of meaning of the spatio-temporal terms can be diagnosed only if we can show that the transition $T \rightarrow T'$ occurs within the extension of a more general idea of space S *that was established already before the advent of* T' [projecting more recent notions back into the past would mean revoking our decision on (a)]. Now I think it will be agreed by everyone that an idea such as S is supposed to be cannot have been part of *common sense* (it would be necessary to assume that common sense is or was able to distinguish between topological, affine, and metrical properties of space and that it is not committed to an unambiguous distinction between spatial and temporal properties). Nor is it possible to locate a suitable S in the *empirical sciences*. Riemann still retained an over-all Euclidian topology. And the contribution of time to the metric did not occur before Einstein. It is of course quite conceivable that S may have been part of some *metaphysical system,* and I for one am prepared to accept a stability of meaning derived from such a background. But, first of all, reference to metaphysics is relevant only if the particular ideas needed are shared by the defenders of T and T'. This is not likely to be the case (absolutism on the part of Newton; relational theories on the part of the forerunners of relativity). Secondly, we may safely assume that metaphysical S's will be rejected by empiricists—and this includes my opponent. Result: the transition $T \rightarrow T'$ involves a change in the meaning of spatiotemporal notions.

This change is drastic enough to exclude the possibility of common elements of meaning between T and T'. To see this, consider the notion of the spatial distance between two simultaneous events, A and B. It may be readily admitted that the transition from T to T' will not lead to new methods for estimating the size of an egg at the grocery store or for measuring the distance between the points of support of a suspension bridge. But in considering (b) we have already decided not to pay attention to any prima facie similarities that might arise at the observational level, but to base our judgment on the principles of the theory only. It may also be admitted that distances that are not too large will still obey the law of Pythagoras. Again we must point out that we are not interested in the empirical regularities we might find in some domain with our imperfect measuring instruments, but in the laws imported into this domain by our theories. Now these laws are very different in T and T'. According to T, the distance (AB) is a *property* of the situation in which A and B occur; it is independent of signal velocities, gravitational fields, and the motion of the observer. An observer can influence (AB) only by actively interfering with either A or B. Any process on the part of the observer that does not reach either A or B leaves the distance unchanged. According to T', (AB) is a projection, into the space-time frame of the observer, of the four-dimensional *interval* $[AB]$. $(AB)_{T'}$ will change even in those cases where a causal influence upon either A or B is excluded in principle. Now one might still wish to retain the idea of a common core of meaning by interpreting the difference between $(AB)_T$ and $(AB)_{T'}$ as being due to the different *assumptions* about "space and time" contained in T and T', respectively. And the locution 'space and time' would now refer to what can be characterized independently of either T or T', though in a manner that conflicts with neither (the last proviso is necessary in order to prevent a return to common sense). Evidently it would correspond to the S mentioned above. We have already shown that no such idea can be assumed to exist. It follows, then, that the difference between $(AB)_T$ and $(AB)_{T'}$ is wholly due to the meanings of the notions used for explaining their properties. In traditional philosophical terminology: $(AB)_T$ and $(AB)_{T'}$ are *constituted* by the basic principles of T and T', *respectively*. These entities cannot be described, not even in part, by means that are independent of either theory at the time of the advent of T'. In earlier papers I have expressed the fact by saying that "$(AB)_T$" and "$(AB)_{T'}$" are *incommensurable notions.*

III

The very same considerations apply if we consider the transition $T =$ classical mechanics $\rightarrow T'' =$ quantum theory (I am now talking about the elementary quantum theory in the form in which it has been developed by von Neumann, and not about the earlier and more intuitive ideas of Rutherford, Bohr, and Sommerfeld). T'' introduces properties whose universalization[4] is possessed by a physical system only if certain conditions are first satisfied. This is true of all so-called "dynamical" properties (angular momentum, which Achinstein discusses is a dynamical property, and so are position, momentum, spin, and so forth; electrical charge is not a dynamical property, at least not within the framework of the elementary theory). Now if we adopt an interpretation of the elementary theory that ascribes this feature to microproperties and macroproperties alike[5] and if we also decide to retain a two-valued logic,[6] we discover again that T and T'' are incommensurable. This feature of the pair has been discussed ever since Bohr introduced the principle of correspondence.

Some authors[7] have commented on the difficulties connected with

[4] If P is a property, then $P \lor \sim P$ will be called its "universalization."

[5] This interpretation is suggested by Temple's proof of isomorphism (*Nature*, 135: 951; cf. also Groenewold, *Physica*, 12: 405 ff). It agrees with our decision on (b).

[6] For reasons, cf. my paper in vol. I of the *Publications of the Salzburg Institute for the Philosophy of Science*, Salzburg, 1965.

[7] Cf. N. R. Hanson, *Patterns of Discovery* (New York: Cambridge, 1958), as well as my comments in *Philosophical Review*, 69, 3 (July, 1960): 251.

a principle that apparently ties together two theories whose concepts cannot be accommodated in a single point of view. This is not the place to examine the matter in detail.[8] Let me only emphasize that the reappearance of conservation laws in the quantum theory cannot be regarded as an argument in favor of a common core of meaning (for such an argument, cf. Achinstein. . . .) For it is clear that the "conservation laws" of the quantum theory share only the name with the corresponding laws of classical physics. They are expressed in terms of Hermitian operators, whereas the classical laws use ordinary functions that always have some value. They allow for "virtual states" which are, strictly speaking, incompatible with conservation. No such states are possible in classical physics. They make use of properties that cannot be universalized simultaneously (position in the potential energy, momentum in the kinetic part) whereas classical properties can always be universalized. Only insufficient analysis could make one believe that the occurrence of so-called conservation laws in the two theories establishes a common core of meaning.

IV

The above considerations, though not sufficient to settle the matter, still provide at least strong prima facie evidence for the existence of "paradoxical" cases in Achinstein's sense. I do not see how the attention to details that Achinstein recommends is going to eliminate these cases. What we need are not further details, but principles such as those involved in our decisions on (a) and (b), which teach us how to evaluate various proposals and which remove the ambiguities characteristic of all semantic information: we can relate the "local grammar" of well-known expressions to different theories in different ways (realism, instrumentalism) and thereby give them different meanings. The adoption of such principles will be guided partly by actual scientific practice, partly by the demands of a reasonable methodology, such as maximum testability.[9] Now, according to both these guides, major revolutions are preferable to small adjustments, since they affect and, thereby, lay open to criticism even the most fundamental assumptions. Achinstein's rule B (which I would formulate as saying that competing theories must have common meanings) puts a restriction on the extent to which we are allowed to revise such assumptions. It does so in the belief that theories can compete only if they are incompatible and that they can be incompatible only if they have common meanings. Is

[8] According to Bohr, the principle of correspondence is a theorem of the quantum theory rather than a "bridge law" connecting quantum mechanics and classical physics.

[9] For details, cf. my "Reply to Critics," which discusses objections by Smart, Putnam, Sellars, forthcoming in vol. II of the *Boston Studies in the Philosophy of Science*.

this transcendental defence of an epistemological conservatism effective. In order to decide the question we return again to our first example, the relation between the general theory of relativity and classical physics.

V

Combine T' with two assumptions which are contrary to fact, namely (i) the over-all metric of space is almost Minkowskian; and (ii) the velocity of light is almost infinite. These two assumptions do not change the semantical properties of T'. T and T' are still incommensurable. Yet it is possible, to a high degree of approximation, to establish an isomorphism between certain selected semantical properties of some (not all) descriptive statements of T' and of some (not all) descriptive statements of T (let the corresponding classes be C and C', respectively). This isomorphism will be valid for finite distances (AB), but not for distances approaching infinity. It will be valid for a finite number of parallel displacements of (AB) around a closed curve, but no longer if this number approaches infinity. Considering that meanings are dependent on structure and not on the particular ways in which the structure is realized, we may say that, within the restrictions given, C and C' have a common core of meaning. We may even identify C and C'. (As $C \neq T$ and $C' \neq T'$, this does not affect the relation between T and T'.) However, C' is formulated in terms of T' and can, therefore, be examined in the manner preferred by Achinstein (meanings shared between the critic and the point of view criticized) within that theory. The examination will of course lead to the rejection of C' and, via the isomorphism, of C, and, as C is part of T, of T also. We see that the show can be rigged in such a manner that the demands for partial stability of meaning are satisfied. But the very method of rigging indicates that the demand is superfluous: when making a comparative evaluation of classical physics and of general relativity we do not compare meanings; we investigate the conditions under which a structural similarity can be obtained. If these conditions are contrary to fact, then the theory that does not contain them supersedes the theory whose structure can be mimicked only if the conditions hold (it is now quite irrelevant in what theory and, therefore, in what terms the conditions are framed). It may well be that those champions of T or of T' who see light only when they see—or believe they see—meanings "are always at least slightly at cross purposes." [10] The fact that argument proceeded even through the most fundamental upheavals, that it was understood, and that it led to re-

---

[10] T. S. Kuhn, *The Structure of Scientific Revolutions* (Chicago: Univ. of Chicago Press, 1962), p. 147.

sults[11] shows that meanings cannot be that essential. I conclude, then, that principle B is neither necessary nor desirable.

VI

These results can be immediately applied to such notorious museum pieces as the mind-body problem, the problem of the existence of the external world, the problem of free will, and to many other problems.[12] In all such cases "new" points of view (which are actually as old as the hills) are criticized because they lead to drastic structural changes of our knowledge and are therefore inaccessible to those whose understanding is tied to certain principles. Now this conservation may well have a physiological foundation. Education, as Prof. Z. Young has put it so well,[13] consists in seriously damaging our central nervous system and in eliminating reactions of which it was initially capable. Admitting such damage and the consequent lack of imagination is one thing. However, one should never go so far as to try to inflict it upon others in the guise of a philosophical dogma.[14]

University of California, Berkeley

[11] The debates during the period of the older quantum theory are an excellent example of discussions of this kind.

[12] For details cf. again my paper referred to in footnote 2.

[13] *Hitchcock Lectures*, University of California in Berkeley, 1964.

[14] For details, cf. the paper referred to in footnote 8.

Selected Bibliography

AYER, A. J., *Language, Truth and Logic*. Oxford: Oxford University Press, 1936. (Also in Dover paperback.) An influential exposition of the positivistic approach to philosophy.

ACHINSTEIN, PETER, *Concepts of Science*. Baltimore: The John Hopkins Press, 1968. A recent, connected survey of the problems discussed in this anthology.

BRAITHWAITE, R. B., *Scientific Explanation*. New York: Cambridge University Press, 1953. (Reprinted as a Harper paperback.) A detailed exposition of the partial interpretation view of theoretical terms and discussion of the consequences of this view for related problems.

FEIGL, HERBERT, and GROVER MAXWELL, eds. *Minnesota Studies in the Philosophy of Science*, Vol. III. Minneapolis: The University of Minnesota Press, 1962. Contains several essays on the problems discussed in this anthology.

HEMPEL, C. G., *Aspects of Scientific Explanation*. New York: The Free Press, 1965. Includes a number of essays on the problems of theoretical terms showing changes in Hempel's position. "The Theoretician's Dilemma" is a particularly clear discussion of the difficulties raised by taking the distinction seriously.

NAGEL, ERNEST, *The Structure of Science*. New York: Harcourt, Brace and World, 1961. Includes chapters which give detailed and influential treatments of the observational-theoretical distinction and its importance.

QUINE, W. V. O., *From a Logical Point of View*. Cambridge: Harvard University Press, 1961. Contains a number of essays on meaning and the connection between language, theory, and reality, including "Two Dogmas of Empiricism," which is the most important criticism of earlier views of language.

SCHEFFLER, ISRAEL, *The Anatomy of Inquiry*. New York: Alfred A. Knopf Company, 1963. A sophisticated attempt to defend a conservative empiricist position.